Family Problems

Oxford General Practice Series 17

PETER R. WILLIAMS
General Practitioner, Oxford

OXFORD NEW YORK TOKYO
OXFORD UNIVERSITY PRESS
1989

Oxford University Press, Walton Street, Oxford OX2 6DP
Oxford New York Toronto
Delhi Bombay Calcutta Madras Karachi
Petaling Jaya Singapore Hong Kong Tokyo
Nairobi Dar es Salaam Cape Town
Melbourne Auckland

and associated companies in
Berlin Ibadan

Oxford is a trade mark of Oxford University Press

Published in the United States
by Oxford University Press, New York

British Library Cataloguing in Publication Data
Williams, Peter R.
Family problems.
1. Medicine. Family Therapy
I. Title
616.89'156
ISBN 0–19–261604–8

Library of Congress Cataloging in Publication Data
Williams, Peter R.
Family problems/Peter R. Williams.
(Oxford general practice series; 17)
Bibliography: Includes index.
1. Family psychotherapy. 2. Family medicine. I. Title.
II. Series: Oxford general practice series; no. 17.
RC488.5.W57 1989 616.89'156—dc20 89-31747
ISBN 0–19–261604–8

Set by CentraCet, Cambridge
Printed in Great Britain by
Bookcraft (Bath) Ltd
Midsomer Norton, Avon

Preface

Tolstoy said of happy families they bore a close resemblance to each other but that every unhappy family was unhappy in its own way.

Whether or not families share patterns of unhappiness, one feature that unhappy families do possess is an instinct to turn to general practitioners for help and advice. The request is often indirect and couched in physical or emotional terms. As a result, many family problems presented in the surgery go unrecognized.

Most families find that coping with illness, loss, and major changes of circumstance is difficult. For some the experience leads to illness and emotional breakdown; others appear more resilient and have less need of professional help. Helping those families in obvious distress and preventing others from experiencing the full impact of disturbance forms an important part of any general practitioner's work and justifies, I believe, a book on family problems.

As well as improving the recognition of family distress, better methods of understanding and working with families are needed which are consistent with the general practitioner's working day. To this end I have selected those ideas and methods commonly used by family and marital therapists and applied them, where possible, to general practice. I have deliberately excluded methods used in dealing with highly dysfunctional families, for example, those in which children are subjected to physical or sexual abuse. It is *ordinary families with problems* rather than *problem families* who form the subject matter for the text, since it is these families who are most amenable to help from a general practitioner with an interest in family pathology.

Oxford P. R. W.
1989

Acknowledgements

Several people have helped me in formulating my ideas about subject matter.

Foremost, were my colleagues at the Kentish Town Health Centre, where I worked before coming to Oxford. I owe a special debt of gratitude to John Horder who has commented on drafts of all the chapters and made many suggestions for improving the text. I also received considerable help from Douglas Woodhouse of the Institute of Marital Studies, and from Peter Tomson, a general practitioner in Abbots Langley with an interest in family work.

Ann Sellers and Gil Dixon provided excellent secretarial help, and the Oxford University Press provided the necessary advice and encouragement to enable the product to reach the bookshelf.

To these and many others, not least family, friends, and colleagues, I offer my thanks.

Foreword

John Horder
Past President, The Royal College of Practitioners

Encouraged by the alternative title 'Family Doctor', general practitioners might easily assume that they recognize the clues in a consultation which point to another family member when the relationship is affecting a patient's health—and that they know through common sense how to respond. They might assume even more easily that they are aware enough when a patient's illness is harming the health of other family members—the opposite situation. Certainly I have met a very few whose awareness of these aspects is obvious, but their work seems to contrast with what the majority of us achieve. Most of us hear and see in the consultation only what our upbringing, training, and experience have led us to notice and consider important.

So are we missing valuable opportunities for helping the people in our care? Is there a challenge here like the challenge of recognizing a depressive illness? There is, after all, a similarity if only because in either case the evidence is so often withheld or presented under some physical disguise.

But, if that is so, how do the family aspects rate in importance compared to all the other aspects of a patient's case which jostle for attention? The generalist's responsibility is broad, but surely there must be limits? Might not consideration of the family be a very low priority?

Bereavement, marital conflict, behaviour problems in children are as likely to force themselves on the doctor's attention as to be concealed. One has no choice but to react in *some* way and there can be few of us who have never felt inadequate in the face of such problems.

Dr Peter Williams has always had a special interest in the family as it impinges on the consultation with one member. His book deals with the obvious unavoidable problems often understood and managed as if they were the problems of an individual. But by outlining the simpler methods which make it possible more easily to see the relevance of others close to the patient, he takes us into less familiar ground. The 'family wheel' and the 'genogram' are tools which may supply information as essential as the record of past illnesses or present occupation. Unlike the auriscope or the electrocardiogram, they may be relevant in every case, but their usefulness can only be tested by applying them routinely.

Least familiar is the simultaneous interview, whether with two members or with a whole family group. The skills are special, but they are not the prerogative of the specialist in family therapy.

Here then is an exploration of territory within the experience of general practitioners every day, but which they may not know or exploit. With its short chapters, emphasis on what is most simple, and concern for the way in which family aspects relate to the wider tasks of diagnosis, treatment, prevention, and self-care, the book can be recommended warmly to any reader who wishes to share the exploration.

Contents

1 Introduction

Family problems are those essentially rooted in the family which find their expression in individual behaviour as well as in relationship difficulties.

A problem may arise in the family as a result of changes, often corresponding to a phase in the life-cycle of either the family or an individual member; or it may be induced by some unforeseen event such as an illness or loss. In the case of illness, the problem is often perpetuated by disordered communication between family members. Other members exercise power over the one who is ill and vulnerable. This vulnerability is sensed by them, often in an unconscious way, and used as an outlet for family tensions which may be caused either by the illness or by some earlier and altogether different conflict.

Solutions to family problems are best sought by bringing about an understanding of what is happening by promoting better communication between family members, and by discussing the rules which govern behaviour in the family so that they are changed.

Family problems are not always presented overtly. They may be suspected when an individual member presents emotional distress or appears to have become more susceptible to physical illness, or when symptoms are ill-defined and defy diagnosis. The general practitioner, who is used to dealing with all types of symptom and illness, and who has knowledge of the family, is usually the first recipient of such problems. Their detection requires skill and intuition. Medical education does not yet equip the general practitioner adequately for this task because he or she is taught to focus on the individual and is therefore liable to forget the role of the larger unit in causing and perpetuating symptoms. It is the family that nurtures individuals when they are young and sustains them through times of stress; it is the family which influences people in the way they perceive themselves as being healthy or sick.

Although we call ourselves 'family physicians' our knowledge of families is often scant. In 1963, Querido showed in his book *The efficiency of medical care*, that doctors are only aware of half the major social problems identified by the families they care for. Rosenberg and Pless (1985), over twenty years later, talked of the paucity of knowledge amongst family doctors about social support, education, and housing. Yet, implicit in the idea of family doctoring is that knowledge of families improves the medical care of individuals. We continue to pay little attention in our records to details of family and social history. What we do remember tends to be limited and anecdotal.

A general practitioner's work demands a **family focus**, or, at the very least, a shift of attention in the direction of close relationships. Above all, he or she must be aware of indirect communication, body language, and significant patterns of speech. A knowledge of the different phases in the family life-cycle is essential, as is an understanding of human behaviour when families are faced by loss, disability, and ill-health.

Once the general practitioner is alert to the family and to relationships within it, as well as to the individual who presents in distress, his understanding of family disorders will inevitably increase. It can be developed by inviting the whole family to participate in problem-solving, by using listening skills, and by actively participating in the family system as it operates in the consulting-room. The doctor must rely not only on what he has learnt in medical school or in the hospital, but on his own experience as a family member. We inherit from our parents more than just a set of genes. We carry forward from childhood a concept of the family based on our relationships with them and theirs with each other. These 'internalized parents' help us shape our view of ourselves as indiviuals, of others, and of the world around us. By implication, understanding our own family life can help us to understand some of the dilemmas our patients find themselves in. Thomas Hobbes spoke of the way we use ourselves as a model for what it is like to be another person, when he wrote:

> Given the similitude of the thoughts and passions of one man to the thoughts and passions of another, whosoever looketh into himself and considereth what he doth, when he does think, opine, reason, hope, fear, etc. and upon what grounds: he shall thereby read and know what are the thoughts and passions of all other men upon the like occasions.

Such reasoning is particularly relevant to the doctor faced with family difficulties in his patients.

In the privacy of the consulting room, doctors are invited by patients to experience the fears and anxieties that go with their symptoms. This can be difficult for the doctor who may feel a lack of skill in responding. If he allows himself to become involved in a cobweb of anxieties, he stands both to gain and to lose. The insight into the ways family members behave towards one another can improve his understanding and lead to better co-operation between him and the patient.

It is a doctor's job to assist in bringing about change; and this is usually difficult. In the case of family problems it is best brought about by enlisting people's co-operation, bringing family members together, and drawing upon the strengths of individual members.

The general practitioner may have the time and skill to do this alone, but also has to be aware that referral may be necessary and be prepared to share his casework with others. He must understand the dificulties that this involves. If he chooses to manage the case himself, he must be prepared to accept limited aims and realize that demonstrating an understanding of a problem can go a long way towards alleviating it. Individuals can be shown that they do have the power to choose alternatives when they feel trapped.

Because of the general practitioner's special role in anticipatory care, his aim should be to pick up family disturbance early and to intervene before patterns become fixed and create pathology. Two settings are particularly helpful to him in this respect—the ante-natal clinic and the child-health clinic; both provide access to people who are receptive to new ideas and who are going through a process of change in any case.

Dealing with family problems can be one of the most challenging areas in primary care; it demands thought, a degree of experience, and the support of colleagues. Knowledge of this area can be extended by reading novels, plays, and other forms of literature which have as their central theme human relationships and the conflicts that surround family life. Literature is a proper and important source of help and because of this I have appended a list of books (Appendix II, p. 101) which I, and others, have found helpful.

OUTLINE OF THE TEXT

Throughout this book I have considered the family as a psychological and biological unit and looked at illness-behaviour in that context.

I thought it wise to begin by examining a few of the many myths that surround family life, as well as providing the reader with some facts about modern families. For example, it is a myth that families care less for their members than they did in the past, with one adult in seven currently providing regular care for a sick, handicapped, or elderly person living in their own home or in another household.

The relationships between illness and family are central to the text. In Chapter 3, the degree to which families cause or perpetuate symptoms in others is discussed with particular reference to schizophrenia, depression, and forms of psychosomatic illness; while in Chapter 4, the effects of illness on the family as a whole are examined.

Since loss, like illness, concerns us intimately as general practitioners, and has a profound effect on families as well as on individuals, it has a chapter of its own. Illness and handicap are themselves losses (of health and function), but loss of a spouse or parent are particular events which carry a high risk of physical and emotional illness.

The particular difficulties involved in recognizing family problems when presented as physical and emotional symptoms form the substance of Chapter 6. Aspects of the doctor's style which influence his ability to elicit important information, are discussed as well as the particular role children play both in creating problems and in signalling family distress.

To help better understanding of family difficulties, three ways of thinking about families are given in Chapter 7: paying attention to the family life-cycle; seeing the family in historical context; and regarding the family as a system.

The practicalities of working with families, from the time the problem first presents to the time when referral may be contemplated, are discussed in the Chapters 8 and 9. The use of 'family wheels' and 'genograms' is described. The particular needs of individuals on their own, of couples, and of parents are separately examined in Chapters 10, 11, and 12.

The need to share or refer casework brings the doctor up against the problems of referral and his relationship with other professionals. Chapter 13 looks at the benefits and difficulties involved in working with other professionals, and I am indebted to Douglas Woodhouse for his help in clarifying my thinking in this potentially troubled area.

Prevention forms a natural sequel to what has gone before, and in Chapter 14 I discuss the vulnerability and resilience encountered in the face of family adversity and the preventive possibilities afforded by crisis, pregnancy, and divorce.

How best to learn and teach family pathology forms the substance of the final chapter, after which there is a list of helping agencies and suggestions for further reading.

2 Family life, myths and realities

Edmund Leach in his 1968 Reith Lectures spoke of family life 'with its narrow privacy and tawdry secrets' as being the major source of man's discontents. Likewise R. D. Laing described families 'with their dirty politics' as being villains of the piece.* Other commentators have shown more enthusiasm for the institution we know as the family, stressing how crucial family relationships are to human dignity and happiness, and how resilient the family is in the face of shifting values.

Family life is certainly a paradox. On the one hand it can act as a cage constraining individual free will and independence; on the other, it provides people with their most intense, uplifting, and memorable experiences. A happy family life still ranks amongst man's highest aspirations—after all, what most people want is a happy marriage, and certainly what children need is a stable and secure family.

We invest more and more in the notion of 'an ideal family', one that is free of problems or that at least resolves its difficulties with ease. The official cult has become that of 'the perfect family' as embodied in the title of one recent book, *The ultimate family: the making of the Royal House of Windsor*.

Yet, for many people the reality of family life is far removed from this image. What is more, people tend to believe that other families have fewer problems than their own, which, as one family therapist has pointed out, is a dangerous myth—destructive because, as their experience inevitably falls short, their best efforts seem to them to be failures.

Much of family life is obscure and private. As doctors we are often only aware of the outer veneer of respectability. When this cracks, as the result of illness or loss, we may become suddenly and painfully aware of problems beneath—of strained relationships, oppression, and exploitation. Because it is so difficult to glimpse, let alone understand, the interior landscape of family life, our impressions are coloured by myths—beliefs which are passed down unchallenged and which can get in the way of understanding. One such myth is that problems only arise because families fail to look after each other as they once did. There are others. If we really believe, for example, that families do not care as much as they used to, it becomes difficult to meet the challenge provided by families who come to us in distress.

* 'The politics of the family are the dirtiest politics of all' is a sentence from Ferdinand Mount's book *The subversive family* (1982), Jonathan Cape, *not* a quotation from Laing, though it might easily be applied to him.

SOME COMMON FAMILY MYTHS

1 'Families care less for their members than they did in the past'

Despite the fact that fewer people now live with their near relatives than previously, contacts between them remain high. Some estimates suggest that between two-thirds and three-quarters of elderly people see their close relatives at least once a week (Abrams 1978; Wenger 1984). Moreover, Willmott in a recent survey (1986) has shown that those people most in need of care receive more visits per week from relatives than those least in need. While in the past the closest family ties appeared to exist only in working-class homes, now as much if not more help is provided to relatives by middle-class families. The help and support people give each other through networks of friends and relatives is overwhelmingly greater than that provided by health and social services. The main carers have always been women. While caring often imposes considerable strain on relationships, help and support are often given by elderly people to their own children in the form of financial assistance, help with child care, and assistance during confinement and illness.

The evidence is strong that families still do the lion's share of caring in our society, from looking after the elderly to caring for the young and handicapped (Allan 1985). Kinship remains a major force in the lives of most people (see Table 2.1).

Table 2.1 *Informal care: the facts*

1. One adult in seven provides informal care
2. Four per cent of adults care for someone living with them; 10 per cent of adults care for people living elsewhere.
3. Four out of five carers* look after someone who is related to them.
4. Carers are found equally among people in different social groups.
5. The peak age for carers is 46 to 64 years.
6. Half of all carers have dependants aged 75 years or over.
7. Many carers who devote long hours to caring have other responsibilities in addition to this role (e.g. dependent children or a job), and a substantial proportion are themselves in poor health.
8. Two-thirds of carers carry the main responsibility for the care of a dependant alone (53 per cent) or jointly with someone else (11 per cent).
9. Women are more likely than men to carry the main responsibility either alone or with secondary support.
10. Only half of all carers have dependants who receive regular visits from health or social services or from voluntary groups.

Source: Hazel Green (1988), *Informal Carers, General Household Survey—1985*, HMSO, London.

* A carer is taken to be someone who looks after or provides some regular service for a sick, handicapped, or elderly person living in their own or in another household.

2 'Marriage is on the decline'

It is paradoxical that at a time when the divorce rate is rising, marriage has never been so popular. It remains true that most young people in Britain will marry, most marriages will survive, most married couples will have children, and most of these children will be brought up by their natural parents (Kiernan 1983). Between two-thirds and three-quarters of marriages are expected to survive as life-long partnerships.

A hundred years ago divorce was infrequent, but there is no evidence that marriage was happier or more fulfilling than now. Indeed, the doctrine preached by the Church concerning the purpose of marriage was that it was not primarily a pleasurable relationship, but rather an institution designed to produce children. Accepting one's lot and devoting oneself to bringing up children were the qualities most stressed. One French physician in a manual concerning the health of married couples gave the following advice: 'Happiness in marriage is not possible unless each keeps perfectly within his role and confines himself to the virtue of his sex without encroaching on the prerogatives of the opposite sex.' (Dr Louis Seraine. *De la santé des gens mariés* (2nd edition 1865), 112–16).

People's expectation of marriage today is altogether greater and may in part account for the rising divorce rate. Nevertheless, considering all the changes that have occurred in this century in the relation between sexes, marriage survives as the institution best able to meet the needs of men, women, and children.

3 'Marriage is a more equal relationship than it was in the past'

If Young and Willmott (1973) are to be believed, we have moved towards a more 'symmetrical' arrangement in which marriage is a partnership in which tasks are more equally shared between husband and wife. There is however little evidence of this. In his study of professional couples, Edgell (1980) found that husbands still have more impact on important decision-making than their wives (see Table 2.2). Most husbands remain peripherically involved in housework and child care. The help they afford their wives in the home is often limited. They remain the main providers of family income, despite the fact that more wives go out to work than before (50 per cent of married women, compared with 10 per cent fifty years ago). Their wives continue to be primarily responsible for things domestic. Much of the work done by women in the home is boring and repetitive and can lead to social isolation and low self-esteem. In contrast, male activities in the home—from do-it-yourself tasks to mowing the lawn—are often perceived as being rewarding in personal terms and likely to gain appreciative comment. While this runs counter to the argument that

Table 2.2 *Impact of husbands on important decision-making in the family*

Decision area	Perceived importance	Frequency	Decision-maker (majority pattern)
Moving	Very important	Infrequent	Husband
Finance	Very important	Infrequent	Husband
Car	Important	Infrequent	Husband
House	Very important	Infrequent	Husband and wife
Children's education	Very important	Infrequent	Husband and wife
Holidays	Important	Infrequent	Husband and wife
Weekends	Not important	Frequent	Husband and wife
Other leisure activities	Not important	Frequent	Husband and wife
Furniture	Not important	Frequent	Husband and wife
Interior decorations	Not important	Infrequent	Wife
Food and other domestic spending	Not important	Infrequent	Wife
Children's clothes	Not important	Frequent	Wife

Source: Edgell, S. (1980), *Middle class couples*: a study of segregation, domination and inequality in marriage, p. 58, Allen & Unwin, London.

contemporary marriage is more symmetrical, Young and Willmott demonstrated equality in the time spent by husbands and wives in tasks which contribute to the home-economy.

4 'A bad marriage is always worse for children than divorce'. Alternatively—'Parents ought to stay together for the sake of the children'.

The death of a parent is acknowledged to be an overwhelming event, but its impact on later behaviour is small compared to the long-term effects of divorce (Van Eerdewegh 1985; Hetherington 1982). With 60 per cent of all divorces involving children, the potential morbidity is huge. On the other hand, conflict within marriage does carry its own legacy of problems for children. It may cause disorder and delinquency in boys, and while the effects on girls are less clear, children—whatever their sex—tune in quickly to parental disharmony.

Richards (1985) points to the quality of parental relationships at the time of the break as being crucial in determining outcome from divorce. There is no doubt that children who have a good relationship with both parents after divorce do better than those who have ceased to have close contact with the non-custodial parent. Where children are subjected to continuing conflict between parents even after the parents have separated, the likelihood of disturbance is great.

The message seems to be clear. As far as is possible, children need to be kept out of marital conflict to preserve their health, and parents should be encouraged to go on parenting after divorce. The context in which the conflict occurs, appears to be more important, as far as children are concerned, than whether or not parents separate or remain together.

5 'The family is itself an outmoded institution'

The difficulties inherent in family life have led some people to question the viability of the family as an institution.

Brigitte and Peter Berger, however, in their book *The war over the family* (1984), point to the lack of agreement as to what the precise nature of the problem is, and what, if any, solutions to it exist. We are witnessing an alarming rise in the rate of divorce and a growing level of domestic violence from non-accidental injury to neglect of elderly dependants. Juvenile crime, violence on the streets, vandalism, and drug abuse have been variously attributed to family disruption and an increasing lack of parental authority. As one newspaper commentator described it: 'Families have come to be regarded as transmission belts of pathology'. Yet institutions are necessary to our survival and there is little to suggest that alternatives to the family are any better or any less problematic. The fact that families do get into trouble and seek help for their difficulties does not mean the institution itself is at fault. Often families lose control temporarily, but in successfully restoring this control regain a sense of dignity and happiness.

The family is rightly celebrated by most professionals as the basic social service. It fulfills an important caring role for which there is no easy substitute. If it is under stress, it is because it needs support rather than wholesale dismantlement.

THE REALITY OF FAMILY LIFE

The starting point for this book is the premise that family life, despite its problems, provides the best solution to the human predicament of wanting close sustaining relationships, a degree of autonomy, a context in which children can be brought up, and a framework within which individual growth can occur.

Family life however, is changing and what is true today may not be true tomorrow. There are fewer large families now than there were. The General Household Survey (1985) shows that there has been a reduction in household size from an average of 2.97 persons in 1971 to 2.56 persons in 1985 (see Table 2.3). This is chiefly due to an increase in the proportion of one-person households and a decrease in the proportion of large households (see Fig. 2.1).

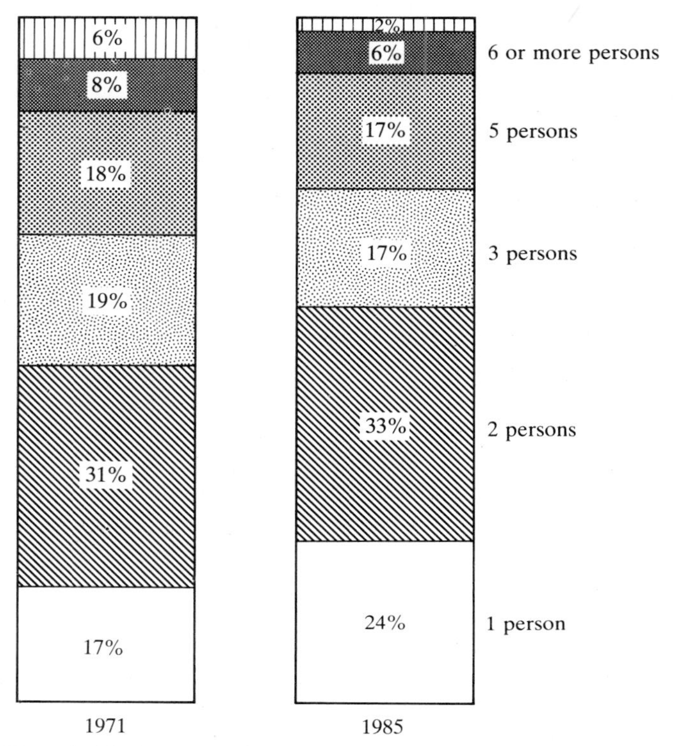

Fig. 2.1 Household size, 1971 and 1985 compared. *General Household Survey* (1985), London, HMSO.

Table 2.3 *Average (mean) household size (Great Britain): 1971–1985*

	Average (mean) household size)	Base number of households
1971	2.91	11 988
1973	2.83	11 651
1975	2.78	12 097
1977	2.71	11 979
1979	2.67	11 490
1981	2.70	12 006
1983	2.64	10 068
1984	2.59	9795
1985	2.56	9993

Source: *General household survey* (1985), HMSO, London.

There is now a wide diversity of family groupings. One-person households account for almost one-quarter of all households but only 10 per cent of all individuals (see Fig. 2.2).

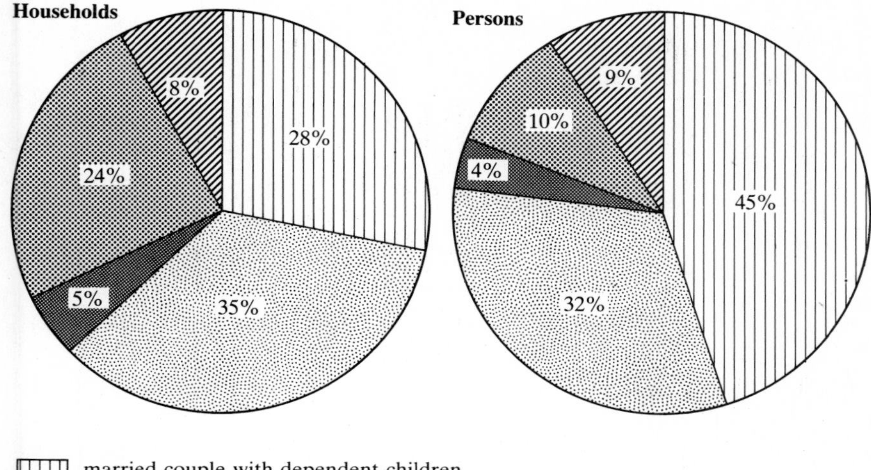

married couple with dependent children
married couple with no children or with non-dependent children
lone parent with dependent or non-dependent children
one person only
other†

* Households categorized by the type of family they contain. In the lone parent and married couple households, other individuals who were not family members may also have been present.

† 'Other' includes households containing two or more unrelated adults and those containing two or more families.

Fig. 2.2 The diversity of family groupings. Households and people by type of household.* *General Household Survey* (1985), London, HMSO.

The number of single-parent families is growing quickly. In 1985, 14 per cent of all families with dependent children were headed by a lone-parent compared with 8 per cent in 1971. This is a consequence of the rising divorce rate (see Fig. 2.3). Although the number of divorces has risen dramatically, the number of remarriages has also increased over the same period (see Fig. 2.4). Many remarried couples go on to have further children, so that many children experience periods in single-parent homes followed by periods in which new relationships need to be forged with step-parents and half-brothers and sisters.

As far as the elderly are concerned, the proportion of those over the age of 75 years is growing (see Table 2.4). Whilst the number of dependent children is not expected to rise over the next thirty-five years, the number of people over pensionable age should increase (see Table 2.5).

All this means a time of adjustment for families—with new division of responsibilities and a greatly more complex set of relationships with which to deal than existed before. For the professionals who work with families, too, it means being aware of the variability and intricacy of family patterns.

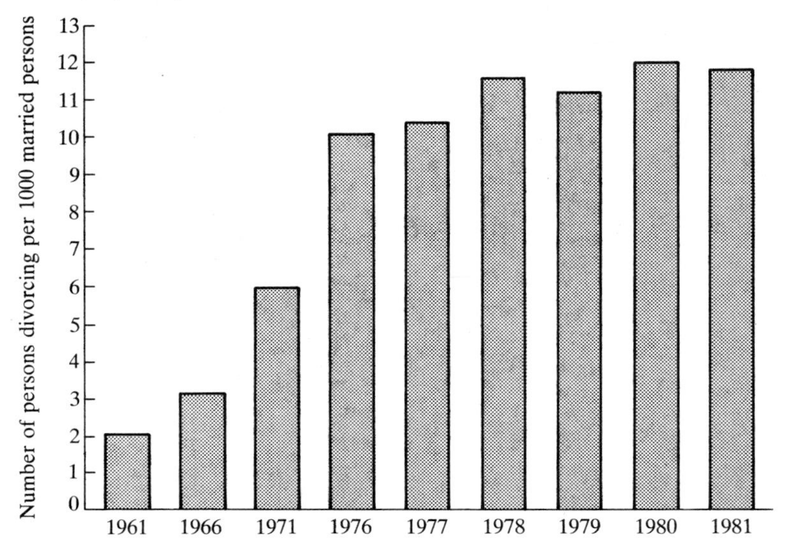

Fig. 2.3 The rise in the divorce rate in the United Kingdom. (In 1971 the new Divorce Act came into operation). *Social Trends* (1983)

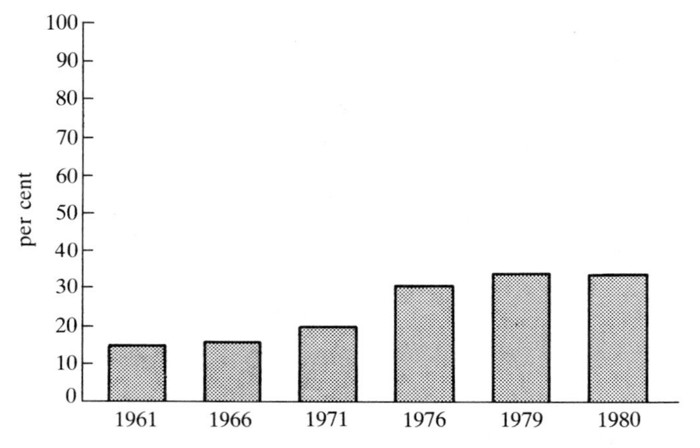

Fig. 2.4 The rise in the re-marriage rate. (Remarriage rate. (Remarriage shown as a percentage of all marriages. *Social Trends* (1983)

If the conventional family of two married parents and two children is no longer the norm in most developed countries, it remains true that a wide range of family types is compatible with normal psycho-social development. Unless the quality of relationships within the family is poor, different forms of family structure appear not to be a major influence on development, although the social circumstances in which people live are of major importance in the development of psychiatric problems. If

Table 2.4 *The growing proportion over the age of 75 years*

Age structure of the population, 1901–2001 England and Wales

Thousands	1901	1951	1971	1981	2001 Projection	Increase per cent 1901–1971	1971–2001	1901–2001
Total population	32 528	43 815	48 854	49 219	51 270	50	4.9	58
65 and over	1 529	4 840	6 531	7 413	7 287	327	11.6	377
75 and over	396	1 581	2 332	2 856	3 314	489	42.1	737
85 and over	50	201	430	514	778	760	80.9	1 456
Proportions								
65 and over/ total population	4.7	11.1	13.4	15.1	14.2			
75 and over/ total population	1.2	3.6	4.8	5.8	6.5			
75 and over/ 65 and over	25.9	32.7	35.7	38.6	45.5			
85 and over/ 65 and over	3.2	4.2	6.6	7.0	10.7			

Source: Rossiter, C. and Wicks, M. (1982).
Crisis or challenge: Family care, elderly people and social policy, London: Study commission on the family.

Table 2.5 *Projected rise in the number of dependants in England and Wales over the next 35 years*

Number of dependants per 100 population of working age*	England and Wales		
	1983	2001	2023
Children			
under age 16	35	36	35
Over pensionable age:			
total	30	29	36
65/60 to 74	20	17	22
75 and over	10	12	14

Source: OPCS Monitor, 27.11.84
Population projections: mid 1983-based.

* Age 16–64 for men –59 for women

families suffer serious financial hardships, are subject to chronic unemployment, or live in unsuitable accommodation, disturbance is likely.

In this book the separate but important influence of illness and loss on family will be considered, since it is in this area that general practitioners have most influence.

3 The family as a cause of illness

SCHIZOPHRENIA

Work on schizophrenia has been of particular value in highlighting the impact families have on the health of their members.

In 1956, Bateson *et al.* published a paper entitled 'Toward a theory of schizophrenia' in which they related schizophrenia to disordered family communication. According to them, impossible demands were placed on a young child by parents who indulged in a highly upsetting form of communication known as the 'double bind'. The child was subjected to pairs of conflicting injunctions or 'binds'. If repeated often enough, and if the child could not escape from these conflicting demands, it was said that he began to doubt the accuracy of his perceptions of other people, opt out of social interaction, and eventually end up a schizophrenic.

Laing, in this country, took up the 'double bind' theory of schizophrenia and developed the idea in his writings. In *The divided self, Sanity, madness and the family*, and in the poetry of *Knots*, Laing provided us with examples of disordered family communication.

For a while the pathology of family communication became an attractive field of interest and many families had their conversations taped and anaylsed by doctors, psychologists, and sociologists. People began to talk of the 'schizophrenegenic family' as a distinct entity.

As an illustration of how families might create illness in their children it sounded plausible, until it was realized that we all grow up with elements of our experience structured by parental double-binds; it is not just schizophrenics who do.

Other forms of disturbed communication in families of schizophrenics subsequently came in for scrutiny, but no-one could be certain that these in any way caused the illness, since they might have arisen as a consequence of living with a psychotic family member.

In a prospective study, Goldstein and Rodnick (1978) set about determining those family characteristics which antedate the onset of schizophrenia. Though deviant styles of parental communication were found to be associated with the development of schizophrenia, the question of how far this reflected heredity rather than environment remained unanswered. Adoption and twin studies have now firmly established the genetic basis of the illness (Kety 1980), though the debate about the interaction between heredity and environment goes on. Certainly no one questions the importance of the family's role once the illness has become established. Brown *et al.* (1972) showed that the family environment into which schizophrenic patients came after discharge from hospital did influence relapse. The best predictor of relapse, in three separate studies, was the amount of 'expressed emotion' in the family. Those with the highest expressed emotion had the highest relapse rate, suggesting that attempts to reduce the degree of emotional involvement of family members with a schizophrenic relative might prove beneficial to the patient. This is the opposite to what might be predicted for depressed patients.

DEPRESSION

Evidence that vulnerability to depression is in part dependent upon family-related factors has come from the work of Brown and associates (1978).

They found that women who had three or more children under four years of age at home, women who failed to have a trusting and confiding relationship with their husband or boyfriend, and women who had lost their own mother before the age of eleven were rendered more susceptible to severe depression than other people. None of these factors alone appeared to be capable of producing significant mood disturbance, but each greatly increased the risk of this condition in the presence of a major life event, which acted as a provoking agent. Brown was able to demonstrate a particular vulnerability to depression in working-class mothers (see Table 3.1). Thirty-one per cent of working-class women compared with 8 per cent of middle-class women developed the condition in response to a major life event. Some of the social class differences arose because working-class mothers experienced numerically more severe life events and major difficulties than their middle class counterparts. However, most of the difference was because working-class mothers were more vulnerable in ways outlined previously.

Brown suggested that vulnerability factors played an important role in limiting women's ability to develop an optimistic view about controlling the world around them, leading to low self-esteem. Anthony Storr linked Brown's ideas to his own psychodynamic thinking by suggesting that what determined whether or not a person had high esteem was a secure childhood relationship with his or her mother.

PSYCHOSOMATIC ILLNESS

Depression is only one outcome of family disturbance. Living in a disturbed family contributes to the severity of psychosomatic illness. Asthma is a good example. Conflict within the family affects the frequency and severity of attacks which in turn heighten the emotional atmosphere at home. The result can be intractable airways obstruction (see Fig. 3.1).

Table 3.1 *Table showing the vulnerability of working-class mothers with children at home (Great Britain), to depression*

Per cent of women developing a psychiatric disorder in a year by whether they have children at home, their social class, and whether they had a provoking agent. Chronic cases are excluded.

	Severe event/ *major difficulty* (%)	No severe event/ *major difficulty* (%)	Total (%)
Women with child at home:			
working-class	31 (21/67)	1 (1/68)	16 (22/135)
middle-class	8 (3/36)	1 (1/80)	3 (4/116)
Women without children at home:			
working-class	10 (3/30)	2 (1/44)	5 (4/74)
middle-class	19 (6/31)	1 (1/63)	7 (7/94)
All women:			
working-class	25 (24/97)	2 (2/112)	12 (26/209)
middle-class	13 (9/67)	1 (2/143)	5 (11/210)
total	20 (33/164)	2 (4/255)	9 (37/419)

Source: Brown, G. W. and Harris. T. (1978), *Social origins of depression*, p. 168, Tavistock Publ.

Understanding the role of the family in this circular system of cause and effect and using a family-based approach to interpret the sequence of events, can have beneficial effects generally. Thus, Lask and Kirk, in a controlled trial of family therapy used as an adjunct to physical treatment, demonstrated improved control in an experimental group of moderately severe asthmatic children. They reported less daily wheezing and had a lowered total gas volume—a measure of air trapping and lung over-inflation. The family therapy which they used aimed at promoting better understanding of the interaction between wheezing and the family system. By improving the skill of the family in coping with an acute attack, and so reducing the anxiety generated in the system, the severity of the asthma had diminished.

A similar approach has been used in treating children with recurrent ketoacidosis. In a study by Minuchin (1974) of 13 diabetic children previously hospitalized on average twelve times a year for severe ketoacidosis, only three had admissions averaging one per annum following family treatment, and ten had no admissions at all on follow-up. Though the numbers are small, the results show a dramatic improvement with therapy.

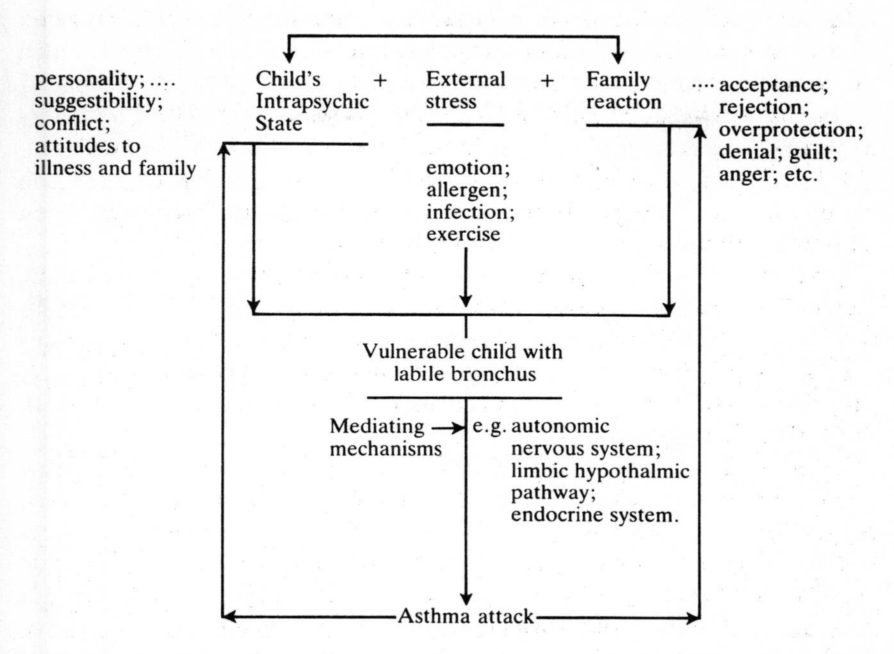

Fig. 3.1 An interactional model for asthma. Lask, B. and Kirk, M. (1979). Childhood asthma: Family therapy as an adjunct to routine management. *J. Fam. Therapy*, **1**, 33–49

INFECTION

Several studies have examined the impact of family stress on the incidence of infection, the most famous one being that of Meyer and Haggerty. In 1962 they examined the effects of acute and chronic family stress on the incidence of streptococcal infection. They looked at 16 families, a hundred people in all, over a twelve-month period, and concluded that infection was four times more likely to occur in the fortnight following acute family stress than in the fortnight preceding it. In those families displaying chronic stress, infection and carrier rates were both increased. The results pointed to family dysfunction as being a factor in lowering people's resistance to infection.

ALCOHOL

Family disturbance has frequently been cited as contributing to drinking problems. Early theories about the development of drinking problems suggested that problem drinkers come from families who were in conflict or who had undergone disruption, but later work (Vaillant 1983) has failed to distinguish families of alcoholics in such terms. What is becoming clearer is that while some kinds of drink problem are related to family problems, others have a strong genetic component. Thus, Goodwin *et al.* (1973) found that there was an association between drink problems in biological parents and drink problems in their offspring, but not between adoptive parents' and children's drink problems.

Whatever the cause, family problems tend to build up when one family member is drinking heavily—behavioural and school problems not uncommonly arise in children, sexual problems and communication problems may develop between husband and wife, and the family as a whole may suffer financial crisis. Excessive drinking becomes a family problem, even if its origins are not familial.

Attempts have been made to influence drinking problems by paying attention to family behaviour, first by helping to change behaviours which trigger drinking, secondly by trying to alter actions of family members which reinforce drinking, and thirdly by teaching the family to provide positive reinforcement and support to the excessive drinker for not drinking. Self-help groups which involve family members of excessive drinkers can offer additional support and instruction in building up alternative life-styles.

Though there are many paths that lead to excessive drinking, some related to family problems, some independent of them, viewing excessive drinking as both an individual and a family problem is essential to any helping strategy.

ANOREXIA NERVOSA

Family relationships have been implicated in bringing on anorexia nervosa and in perpetuating it, but the evidence is largely anecdotal. Families of anorectic patients often have a greater than average preoccupation with weight, eating, health, and exercise, which antedates the onset of the disorder (Kalucy *et al.* 1977). Many families include doctors, dieticians, cooks, and waiters. But a more important predisposing influence may be a lack of ability on the family's part to deal with the problems of adolescence. Crisp *et al.* (1974) found that restoring an anorectic daughter's weight by inpatient treatment resulted in an increase in psychopathology amongst parents, suggesting that the daughter's illness had some protective function as far as the parents were concerned. The anorectic child may play an important role in maintaining family homeostasis.

PSYCHOSOMATIC FAMILIES

Any illness can be influenced dramatically by family attitudes and behaviour. Although it is often possible to see the role of the family in maintaining the disease process, it would be wrong to apportion blame. To use parents as scapegoats, for example, in cases of childhood disorders, is unhelpful, though their role in perpetuating the problem may need recognizing.

It was Minuchin *et al.* (1978), once more, who drew doctors' attention to the way in which children from certain families had their symptoms sustained by their parents. Such children appeared particularly vulnerable to illness, had parents who were in constant conflict, and belonged to families who shared certain other characteristics. Four features in particular appeared important in their families.

1 Enmeshment

Changes in one family member, or in the relationship between two members, had a profound effect throughout the family. Boundaries that defined individual autonomy were weak, making individuals vulnerable to changes affecting others in the family.

2 Over-protectiveness

Family members showed a high degree of concern for each other's welfare. In particular children whose physical symptoms had an important psychological component, felt a great sense of responsibility for protecting the family.

3 Rigidity

Family members found it difficult to accommodate to change, indeed on occasions they denied the need for any change.

4 Inability to resolve conflict in a satisfactory way

Some members argued interminably without achieving any result; others avoided discussing problems; some simply denied the existence of any problem whatsoever.

Minuchin saw the child's illness as playing a vital role in family conflict. This could occur in one of three ways

1 Triangulation

Each parent demanded that the child side with him against the other parent. By siding with one, the child was automatically seen as attacking the other.

2 Detouring

The parents submerged their own problems in problems of parenting. Either the child was defined as 'sick' and in need of their collective help, or as 'bad' and the cause of that family's problems.

3 Forming a coalition

One parent joined the child to form a stable coalition against the other parent.

He suggested that these patterns were used again and again by families operating under conditions of stress.

Minuchin's ideas receive support from the observation that it is the avoidance of open argument that creates most stress in families, not the rows themselves.

Though it is often difficult to arrive at firm conclusions concerning the family's role in causing or maintaining symptoms, much individual behaviour appears inexplicable and even socially senseless without a family framework. Realizing that families can influence the course of illness in a positive as well as negative sense forms the basis of much of the therapy outlined in later chapters.

4 The effects of illness on the family

Every illness has a family component. The impact of the ill-health falls not just on the individual who bears symptoms, but on those who live with the sick person and who attend to his or her needs. Illness is often a particularly stressful time for the family. The bonds between the ill person and other family members can become weakened and disrupted resulting in 'family crisis'. In this situation it is not uncommon for the general practitioner to find 'clustering', with problems like depression and anxiety occurring simultaneously in several family members. In other families, individuals present with different physical complaints—headaches, neck pain, gastro-intestinal disturbance, for example—as though vying with each other for medical attention. Such grouping of problems and aggregation in consultations amongst members of the same or related families should alert the doctor to a state of 'family crisis' affecting more than the person in the consulting room.

CHRONIC ILLNESS AND HANDICAP

Sudden unexpected illness carries its own immediate threat of disruption, but chronic illness and disability, by its unremitting nature, can have an even more profound effect on families.

Mental illness is arguably the most difficult type of illness for families to deal with. Yarrow *et al.* (1955) found that wives of patients diagnosed as having a mental illness often resisted the definition of their husband's condition. Their fear was that the family would be stigmatized. The conflict between needing help and not being able to receive it understandably created problems for them in their attempts to hide the illness from others.

In a study of the impact of mental disturbance on the family, Reading (1986) found that in over half the families studied there was a sense of personal despair or profound loss. This was tied to a sense of failure because the family was unable to bring about the recovery of the sick member. The cost to the family of having a mentally sick relative could not be ignored. For low-income families this was found to be significant, though less so for professional and middle-class families. Needless to say, restrictions in families' life-styles were common with day-to-day routine being dominated by the illness. A particular worry in some families was the possibility of others in the immediate or extended family group becoming ill. The effect on marital relationships was surprising, with a tendency for this to get either markedly worse or significantly better. One husband commented that he and his wife came closer together and now talked to each other more. Another couple said that understanding, insight, and mutual support had increased. They even described how a forgotten shared sense of humour had returned and how this had helped them to respond to their mentally-ill son with less intensity. Other couples talked of serious marital difficulties and sexual problems.

Reading found, like Yarrow, that there was a tendency for families with a mentally ill member to close ranks in an attempt to keep the illness secret. Thus, the affected person was sometimes described as 'working away from home'. In some families the actual description of the illness was altered to what was considered more acceptable to the wider family or community.

It has come to be recognized that having a child with a chronic condition like epilepsy can, like having a mentally sick family member, create considerable stress for parents and other children in the family—so much so that Hoare (1984) urges all doctors to consider not just the needs of the particular child with epilepsy, but those of the parents who need advice and counselling from the start. Parents of epileptic children, he found, were often frightened by their children's fits, had little knowledge of the underlying condition, and were frequently unsure what attitude to adopt towards their children. The children themselves tended also to have more behavioural difficulties than other children. Exaggerated dependency was often an additional burden for the family to cope with—hence the importance of minimizing restrictions on activity and stressing education not just of parents but of school teachers too.

Adults with handicapping conditions pose as many problems for their immediate family as children do. Carnwath and Johnson (1987) examined the various disadvantages suffered by families of patients with a stroke. Since stroke accounts for a quarter of all severe handicap in the community, its consequences for those who care for stroke patients are potentially great. They found that the spouses of stroke victims were particularly vulnerable to depression. Moreover depression was likely to be present for a considerable time after the stroke had occurred. Looking after a stroke patient often caused the spouse to give up work and this could result in financial difficulties. The relationship between man and wife often altered radically. Personality changes, problems of communication, and emotional lability all played a part. Resentment on the part of the carer could set in along with feelings of guilt and depression. Depression in the spouse would often militate against successful rehabilitation of the stroke patient. Carnwath found that few depressed spouses were given anti-depressants though many took minor tranquillizers, despite a lack of evidence that benzodiazepines were effective in depression. The implication of the study was that proper attention paid to the spouses of patients with stroke would not only benefit them but improve the prospects for both.

ACUTE ILLNESS

Few studies have examined ways in which more acute illness can affect family life. Johnston *et al.* (1985) studied the impact of whooping cough both on the child and on the parents and children who shared the family home. Understandably, most children were very distressed by their symptoms, but the illness also proved an enormous strain for their parents and siblings. Apart from universal and extensive sleep disturbance, several families experienced marital problems and, amongst the siblings, most complained of feeling tired, of being neglected by their parents, and of feeling jealous of the attention given to the sick child.

The psychological and social disability in the wives of men suffering a first acute heart attack was also found to be considerable in a study by Mayou *et al.* (1978). The disturbance was not only persistent, but comparable with that of the patients themselves. As in the case of strokes, the spouse has a major role in encouraging the patient to get better and the implication is that families need greater practical support, and that advice needs to be given to the spouses of coronary victims on how to cope with their feelings.

HANDICAP IN CHILDREN

Many studies suggest that the families of handicapped children experience more stress and less psychological well-being than other families.

Reaction of parents to a handicapped child appears to be influenced by whether or not the handicap is present from birth, by its severity, by the attitudes of others, and by the prevailing culture. That the presence of a handicapped child can also adversely affect the psychological health of other siblings is clearly shown in a study by Gath (1978) on school-age siblings of Down's syndrome children. MacKeith (1973) lists some of the mixed feelings shown towards a Down's baby. The reaction of doctors was as mixed as those of the parents and was reflected in different ways (see Table 4.1). Support is needed especially in the early days if families are fully to accept their role as providers of care and not become ill themselves. There is need, as Bax (1978) says, for doctors and health visitors to see families of handicapped children on their own home ground. Stress from repeated hospital admissions, visits to the GP surgery, as well as the actual nursing burden, conspire to make a normal social life for many of these

Table 4.1

Reactions of parents to a handicapped child

1. Biological reactions:
    (a) Protection of the helpless.
    (b) Revulsion at the abnormal.
2. Feelings of inadequacy about:
    (a) Reproduction.
    (b) Rearing.
3. Feelings of bereavement (at the loss of the normal child they expected):
    (a) Anger.
    (b) Guilt
    (c) Adjustment.
4. Feeling of shock.
5. Feeling of guilt.
6. Feeling of embarrassment (a social reaction to what the parents think other people are feeling).

Reaction of their professional advisers

1. Feelings of revulsion at the abnormal (they may recommend that the parents put the child away into a home).
2. An over-solicitous response towards the child, so that the needs of the rest of the family are forgotten.
3. Feelings of inadequacy (may be shown by brusque dismissal of the child and parents or, paradoxically, by an objection to the parents 'shopping around' for a further opinion).

Source: MacKeith, R. (1973). *Developmental medicine and child neurology*, **15**, 524–5.

families impossible. It is interesting to note however that not all families experience the same degree of stress, and as much appears to depend on the characteristics of the families themselves as on the specific disabilities of the child.

THE CHALLENGE OF ILLNESS

Most sick people are given special consideration by others and their behaviour is excused in ways that would not occur had they been well. The challenge of illness lies as much in the ability of people to cope with sickness in others as in the patient marshalling his powers of recovery. The sick, like the very young, the dependent elderly, and the handicapped, all impose a strain on family life and constitute a test of family relationships. Not only does illness on occasions threaten family stability, it challenges the very core of family life; its sense of privacy. Strong feelings which normally lie dormant can surface under the strain of caring for a sick family member and create a tense emotional climate in the home.

One particular consequence that illness (real or imagined) has on the family is its effects on the power structure. The sick member is often granted special licence by the family—his behaviour is somehow excused. On occasion this results in the sick person being invested with great power, the exercise of which over the other family members is liable to become a problem.

It is worth reflecting on the potential effects on family members of a sick but powerful relative. All too often it is assumed that the patient is powerless and that control is exerted only by relatives. In practice, the balance of power can shift in either direction. Helping those families who become locked in a power struggle can be difficult since solutions which involve relinquishing power or coming to terms with the situation can be as painful as the illness itself.

CARING FOR OTHERS

Recently the important role that relatives play in caring for the ill and the handicapped in the community, and the heavy burden that families carry, has been increasingly acknowledged. Most research has focused on care for elderly dependent relatives and for adults with chronic illness. It has become clear that one person in the family, usually the spouse or a close female relative, offers most care. Mood disturbance amongst carers is common and not always recognized by doctors. Even those who on the outside cope with repetitive nursing tasks and disruption of social life pay a price in terms of exhaustion and of stress in their marriage.

Anderson (1987) lists some of the needs of carers.

- Recognition of their work;
- Planned respite care;
- Information about disabilities and services;
- Physical help;
- Money;
- Continuity of support to help them respond to changing circumstances and prevent feelings of abandonment.

Where the carer is no longer able or willing to cope, he or she should feel confident that alternative institutional care of high quality is available.

Anderson suggests that the primary care team should become the advocates of the carers and make themselves accessible to the needs of the family. Sadly the support provided to carers is woefully inadequate. It is wrong to think that what most relatives of a sick person want is more hospitalization. It is support and information in providing care in the patient's own home that families desire.

5 Loss and the family

Loss, whether threatened or real, has a profound effect on family life. The ripples of disturbance may not confine themselves to the immediate family but affect friends, neighbours, and occasionally the community as a whole.

Loss does not have to result from death to cause upset. Already we have seen how loss of health and bodily-function can have serious consequences for the rest of the family as well as for the individual who suffers illness and disability. Likewise, the loss of a job can be a serious blow to family morale, resulting in financial hardship and affecting the self-esteem of the person made unemployed. Loss of a spouse through divorce or desertion carries its own particular legacy of disturbance, especially where there are children involved. Divorce nearly always causes profound sadness and in many cases results in frank illness and behavioural disturbance. With the median time for marriage standing now at seven years, the number of adults and children involved in loss through divorce is enormous.

Any loss necessitates adjustments, often of a major sort, and where these prove difficult, family problems can be anticipated. The social re-adjustments involved in various life changes have been examined by Holmes and Rahe (1967). The impact of certain life events, many of them losses, on health, led the authors to construct a rating scale with each life event accorded a value as well as position on the scale (see Table 5.1). It can be seen that losses experienced within the family setting ranked high on the social readjustment scale. Loss of a spouse through death, divorce, or separation came out top, closely followed by the death of a close family member.

LOSS OF A SPOUSE

Much has been written about the health consequences of bereavement. Cross-sectional studies point to a dramatic increase in mortality amongst those recently widowed, the mortality rates rising on occasions ten times above what would be expected for a particular age group (Kraus and Lilienfeld, 1959). A smaller increase in mortality was noted by Parkes *et al.* (1969) in his prospective study of widowers. This increase was accounted for chiefly by heart disease in the first six months of a bereavement. By controlling for age, sex, and residence over a ten-year period, Helsing *et al.* (1981) confirmed this increase in mortality—but this, it seemed at first, was as true for widows as for widowers. By taking into

Table 5.1 *List of life events arranged in descending order of impact on health*

1	Marriage
2	Troubles with employer
3	Detention in jail or other institution
4	Death of spouse
5	Major changes in sleeping habits (a lot more or less sleep, or change in part of day when asleep)
6	Death of a close family member
7	Major change in eating habits (a lot more or less food intake, or very different meal hours or surroundings)
8	Foreclosure on a mortgage or loan
9	Revision of personal habits (dress, manners, associations, etc.)
10	Death of a close friend
11	Minor violations of the law (e.g. traffic tickets, disturbing the peace, etc.)
12	Outstanding personal achievement
13	Pregnancy
14	Major change in the health or behaviour of a family member
15	Sexual difficulties
16	In-law trouble
17	Major change in number of family get-togethers (e.g. a lot more or less than usual)
18	Major change in financial state (e.g. a lot worse off or better off than usual)
19	Gaining a new family member (e.g. through birth, adoption, elderly relative moving in, etc.)
20	Change in residence
21	Son or daughter leaving home (e.g. marriage, attending college, etc.)
22	Marital separation from partner
23	Major change in church activities (e.g. a lot more or less than usual)
24	Marital reconciliation with partner
25	Being fired from work
26	Divorce
27	Changing to a different line of work
28	Major change in number of arguments with spouse (e.g. either a lot more or less than usual, regarding child-rearing, personal habits, etc.)
29	Major change in responsibilities at work (e.g. promotion, domotion, lateral transfer)
30	Wife beginning or ceasing work outside the home

Adapted from: Holmes, T. H. and Rahe, R. H. (1967). *Journal of Psychosomatic Research*, Vol. II, pp. 213–18. Pergamon Press.

account other variables such as social class and smoking habits however, his results become significant only for widowers, with remarriage apparently reducing the overall mortality rate.

The adverse effects on health of losing a spouse cannot be denied, but it is debatable which of four explanations explains the increase in mortality amongst widowers.

The explanations put forward are as follows:

(a) Those who are 'unhealthy' tend to marry those who are themselves 'unhealthy'

In other words, people with physical disability, or who have habits which are likely to affect their health such as smoking or being obese, chose others with similar health problems. The result is a shared tendency to die prematurely or at a similar age. To support this theory, the Framlingham study on coronary risk showed a higher concordance between spouses of factors such as blood pressure, cigarette smoking, elevated cholesterol, and excess weight than would be expected by chance. Moreover, this did not change with time, implying it had more to do with the selection of partners than sharing the same environment.

(b) Couples tend to share an unfavourable physical environment and so are subject to the same amount of risk

They may live in a polluted atmosphere, be close to toxic wastes, or eat food which is high in saturated fats and low in fibre, thereby succumbing to the same illnesses.

(c) Loss of a spouse leads to loss of care, which leads to premature death

The web of social ties that surrounds an individual often weakens after the death of a spouse. Moreover those in poor health tend to possess fewer social contacts and systems of support than those in good health. The fact that loss of a spouse can have a persistent effect on health may point to loss of social support as being the most harmful factor as far as physical and emotional health is concerned.

(d) Loss of a spouse leads to a decline in resistance to illness:

Schleifer *et al.* (1983) found that there is a significant decline in lymphocyte funtioning during bereavement, which lends support to the argument that after the death of a spouse there is a decline in immunity. This will lower resistance to illness. The frequency with which the bereaved neglect nutrition and even personal hygiene adds substantially to the risk of illness and might explain the excess in mortality amongst men who previously relied on their wives to cook and shop for them.

One late sequel of bereavement is the **anniversary reaction**. Sometimes several years after the loss, and on the precise day the person died, physical symptoms often mimicking the illness which caused the death, can arise unexpectedly. While some are clearly conversion reactions, others are organic in nature, for example heart attacks, relapses in ulcerative colitis, and profound mood disturbance. The anniversary, it has been argued, is a time of unconscious psychic stress which activates symptoms in the individual reflecting those experienced by the dying person in his last illness (Pollock 1970). Such reactions may be commoner in practice than is at first realized. Having an eye to anniversaries and pointing out the temporal link between symptoms helps the bereaved person to understand and to come to terms with his current condition; it may prevent a recurrence of symptoms in the future.

LOSS OF A PARENT AND A PARTNER

The effect on a child of losing a parent as a result of death, desertion, or divorce, like the effect on the remaining partner, is bound up with the quality of the relationship before the break, as well as the home circumstances before and after the loss.

While children who are brought up in homes with just one parent are more likely to have behavioural problems, the extent to which problems stem from the lack of a parent rather than other adverse family features is difficult to assess (Rutter 1975). Thus, single-parent households tend to have a lower income than two-parent homes. The Finer Report (1974) estimated that half of all fatherless families relied on supplementary benefit for their income in the early seventies. There is little to suggest that this stage of affairs has improved. Employment of the mother offers only a partial solution to this problem, pay being often too low to provide adequately for the rest of the family. In 1981 just under 50 per cent of families receiving Family Income Supplement were single parent families, and many are still close to the poverty line. One-parent families additionally face other disadvantages—poor housing and discrimination both in the job market and socially in the community in which they live. One-parent families often feel isolated. Depressive illness amongst one-parent mothers is higher than in women from unbroken homes. Parents tend to be anxious and self-critical. The model of parenting with which the child grows up is often distorted. Remarriage does not guarantee satisfactory parenting and reconstituted families have their own problems.

Of the many children who are involved in family break-up, a fifth at least are affected adversely in ways which can reach the general practitioner. The disturbance can be difficult to spot, disguised as it may be in various physical forms, but it is important to realise that early intervention—even before the break has finally occurred—can result in lasting improvement to the mental health of offspring.

Compared to the effects of loss of a parent through death, the long-term effects of divorce on children's behaviour are much greater. The context in which the separation occurs and the meaning which this has for all family members appears to have more influence than the break itself. Establishing a good relationship with both parents after a divorce seems to protect the child against later disturbance, while maintaining a state of conflict predisposes to later problems.

LOSS OF EMPLOYMENT

The impact of unemployment on families has not been formally studied despite its potential for causing upset. Being unemployed carries with it a stigma not only for the person out of work but for the rest of the family. Most unemployed are poor and their families suffer considerable hardship. Anxiety, depression, and other nervous symptoms are especially prevalent amongst the families of the unemployed. There exist higher levels of marital stress and breakdown in these households with all the adverse effects on children previously mentioned.

The relationship between unemployment and physical health is far from clear. Controlling for age and social class, White (1983) found that unemployed men were only slightly more susceptible to chronic illness than those with a job. Other studies, however, point to an increase of up to 20 per cent in mortality amongst male unemployed. Male family members, in particular, commonly experience a sense of profound personal failure and inadequacy which is not helped by the prevailing attitude that the unemployed are scrounging off the state.

Work is important in the development of personal self-esteem. As Galen pointed out, it is nature's best physician. As well as giving structure to the day it provides status for the family, financial security, and meaning to leisure activities; but not all work is satisfactory. Many jobs can be a source of stress which spills over into family disturbance. Overall, however, work tends to be beneficial to the individual and his family. Reactions to unemployment are clearly influenced by the personality of that person, the nature of the previous employment, the time spent out of work, and by susceptibility to stress in general. Those most at risk from psychological reactions would appear to be the young, the unskilled, people from abroad, and the disabled. The families of such people bear the brunt of such disturbance.

RESILIENCE AND SUSCEPTIBILITY TO LOSS

Though loss is unequivocally linked to disturbance, it is only a minority of individuals and families who suffer permanent disability. General practitioners are made only too aware of differences in response to loss. Some families appear to succumb readily to upset of any sort; others surprise us by their resilience. This variation in response has led researchers to look for factors that increase or decrease susceptibility to loss. Much attention has been devoted to the level of support that operates within the family and from extended family and friends. Having a close-knit family group who are prepared to lend a hand or offer a sympathetic ear helps, but if relationships are already strained, loss can be a factor that leads to further estrangement rather than support.

For adults, having a satisfying marital relationship is beneficial since it enables people to cope better with any loss. The number of close relations a person has is of less importance than a person's satisfaction with the relationships, even if these amount to only one or two. In children, the presence of a secure relationship with one parent often compensates for an insecure relationship with the other. Very young infants are to some extent protected from the ill-effects of loss since they are relatively non-selective in their attachment to people. This is not, of course, true of older children. Having stable early attachments and possessing a high sense of personal self-esteem is thought to breed resilience to loss, perhaps by creating a feeling of control over situations which otherwise threatens to undermine self-confidence. Instead of merely reacting to loss such people feel compelled to act positively. Finally, the mere experience of loss and of coping with it certainly helps individuals and families when faced with a similar situation again.

Any loss leads to some form of grief. Making the connection between loss and grief and giving permission for grieving to occur should form part of our professional work. By being available, by listening in a non-judgmental way, and by putting a new perspective to the grief process, we can help the family realize that significant losses provide a tragic opportunity for growth.

6 The recognition of family problems

The general practitioner is trained to give his patients the opportunity to present a wide variety of problems. Many can adequately be dealt with by himself. Family problems are no exception but, instead of being presented clearly and explicitly, they are often couched in physical ways such as headache and lassitude, or in emotional ways such as the expression of anxiety and depression.

Because the symptoms of family distress are very often physical, they can easily be mistaken for those of organic illness, or dismissed entirely as minor and not worth the expenditure of time. Even when there is a hint of understanding between doctor and patient, both parties may collude in keeping the lid of Pandora's box tightly shut for fear of letting out more than either can comfortably handle. It is safer on occasions to talk of organic illness than deal with family distress.

People often feel ashamed when they cannot adequately handle the difficulties that arise in their relationships with others. It is easier to talk of chronic neck pain than say that a marriage is failing, or of a fear that one's blood pressure has risen than admit that an elderly relative is causing anguish to the rest of the family by refusing help in the home.

It is with some temerity that some patients come seeking help from their doctor and many are deterred altogether by a cool reception, insufficient opportunity to talk, or a consultation which is brought to a speedy conclusion with the prescription of a drug. Some many feel emboldened to ask the doctor about their problems at the end as a 'by the way . . . consultation' at the door, but many remain silent.

Szasz (1961) once remarked that whenever the relationship between two people is uncertain, and hence when either communicant or both feel threatened and inhibited, the stage is set for the exchange of **indirect messages**. These fulfil a dual function—first to transmit a degree of information and secondly, to explore and modify the nature of the relationship. Doctors need to be alert to indirect messages in the same way that they look for the metaphorical meaning behind patient's statements—the painful neck which may signify a nagging constraint in that person's life or the difficulty swallowing which has no organic basis and may point to a behaviour in someone close which cannot be 'stomached'.

A case which illustrates some of the difficulties of recognizing family problems is that of the Carter family.

Caroline Carter wrote a note to her doctor requesting sleeping pills.

Her nine-month old son, John, was demanding of her time day and night and Caroline felt tired and irritable in the day-time. Her husband, Martin, was unable to offer much help as he worked nights. Some time later Caroline herself presented in the surgery complaining of difficulty swallowing. 'I'm afraid to,' she said, 'every mouthful is agony, and yet I'd give anything to eat.' The symptoms had been getting worse despite the fact that her weight was constant and her periods regular. Investigations, including a cine film of swallowing, were normal. Only later was it discovered that the onset of her symptom corresponded to a row she had had with her husband when he had discovered an affair she was having with another man—the best man at their wedding. Although there had been a reconciliation, their relationship was far from easy. It transpired that he, too, had been requesting sleeping pills from his doctor, and their daughter, Julia, had been investigated for persistent and inexplicable chest pains. It seemed that Mr Carter had become jealous of Julie's relationship with her mother and the two argued incessantly. The previously 'unaffected' member of the family, John, had by now developed asthma and saw to it that he, too, received attention from family and doctors alike.

POINTERS TO RECOGNITION

Not all family problems are so indirect in their presentation, but it is useful to consider what pointers there are to their recognition in the consultation. There are six common ones:

1. Vague symptomatology which does not fit into any clear medical model or diagnosis. Common examples are tiredness, joint pains, and dizziness.
2. Frequent consultations over apparently 'minor' illness.
3. Emotional problems, notably depression.
4. Addictive behaviour, for example an increase in the consumption of cigarettes or alcohol, shown as frequent attacks of bronchitis or a request for a sick note and a bottle of antacid on Mondays.
5. Management problems and behavioural difficulties in children, from sleep problems to temper tantrums.
6. Loss of control if there are chronic conditions like asthma, diabetes, and epilepsy.

While it is not true that every person who presents with lassitude should automatically be suspected of having a family problem, the assumption that it is invariably organic and merits a blood count and E.S.R. needs challenging. The yield of positive results is low in such a case and the emphasis on the laboratory may be an expensive diversion. Thinking in

Table 6.1 *Ten medical behaviours which relate to accuracy of psychological assessment and which can be significantly improved by training.*

Start of interviews:

1 Making eye contact with the patient.
2 Clarification of the presenting complaint.

Form of the questioning (see text):

3 Proportion directive (rather than closed).
4 Use of directive questions when dealing with physical symptoms.
5 'Open to closed cones'.
6 Focusing on the present rather than the past.

Some special techniques:

7 Sensitivity to verbal cues relating to psychological distress.
8 Sensitivity to non-verbal cues indicating distress.
9 Ability to deal with over-talkative patients.
10 Ability to handle the notes.

Source: Goldberg, D. (1982) *Psychiatry and general practice*. ed. A. W. Clare and M. Lader pg. 40, Academic Press Orlando, Florida

family terms, and exploring the effect of the tiredness on that person's life and the lives of those close to him can often be a more rewarding avenue.

DOCTOR BEHAVIOUR

In recognizing family problems, as in recognizing psychiatric illness, the doctor with the emphathic manner who enquires about his patient's home life is better able to make an accurate assessment than the doctor who ignores this approach.

Goldberg (1982) lists ten behaviours which he finds closely relate to a doctor's ability to make accurate diagnostic assessments (Table 6.1). In the list, a 'directive' question is one which indicates the aspect of the patient's complaint which interests the doctor. Unlike a 'closed' question it cannot be answered by a simple 'yes' or 'no'. Goldberg found that doctors who were less adept at making psychological assessments asked several closed questions and tended to ask them early in the sequence of questioning, preventing their patients from expressing their problems in their own words. 'Open to closed cones' relates to the sequence of a doctor's questions. Ideally, a doctor should begin by asking 'open' question and when he has heard the general nature of the presenting complaint, should follow this with 'directive' questions.

The importance of Goldberg's work lies in the ability to train doctors in the technique of interviewing so that their performance in recognizing psychological (and therefore often family) problems is improved.

To convince oneself that a patient's symptoms are both real and have their roots in some relationship difficulty within the family can take time and requires a lot of tact. Physical examination can be as important as verbal enquiry into relationships, and the simple act of laying on hands can be enabling to patients who find it difficult to express themselves clearly. Even then, when the complaint has been treated seriously and when sufficient opportunity has been given to the patient to indicate whether or not there are family problems, the patient may be reluctant to divulge what is going on at home. Appealing for more information from family members close to the patient can help to bring things into the open where suspicions run high. Once the patient's permission is obtained, few relatives can resist an invitation along the following lines:

> I am concerned about your husband/wife/child's health and before proceeding further would very much value your ideas as to what is wrong and what could be done to help. Could you possibly come along and discuss the matter along with the patient?
>
> Yours sincerely

Engaging families in recognizing that their problems lie not with one individual but several, is a skill which needs practice for it to come easily. Once mastered, and given a willingness on the family's part to accept that they are involved, the stage is set for intervention.

THE ROLE OF CHILDREN

Turning to the role of children as pointers to family disharmony, it is frequently the case that children, identified as being patients, are the bearers of symptoms for the rest of the family. Consider the case of the Conrad children:

Mr and Mrs Conrad and their son, Mark, and daughter, Fiona, had recently moved to London where Mr Conrad had taken up an appointment as a systems analyst for a large city firm. Wasting little time, they had referred themselves to a hospital specializing in paediatric disorders.

Their concern was chiefly about Mark, a somewhat withdrawn child of five, given to outbursts of rage and occasional fainting spells which had so far defied diagnosis. From the start Mark had appeared difficult—even his birth had been traumatic—and Fiona, two years older, was showing signs of jealousy at the attention that Mark was getting from his parents.

Husband and wife found it difficult to see eye-to-eye about their children's management and the disagreements had imposed a considerable strain on the marriage. There had been freqeuent rows with parents taking opposite views over their son's behaviour: Mrs Conrad believed Mark had an undiagnosed neurological disorder, while Mr Conrad felt his son was putting on a display for his parents (notably his mother) to get more of their attention. Fiona caused her parents concern, too, as she had episodes of adominal pain and headache which made her lose a significant amount of time from her schooling.

Contact with their doctor before the move to London had been made exclusively in connection with the children. Mark, on the whole, was identified as having most problems while his sister had been the recipient of occasional night visits by emergency doctors and had on one occasion been admitted to hospital overnight with suspected appendicitis. Mark had seen two specialists in the North about his 'funny turns' and neither had reached any firm conclusion, although one had raised the spectre of temporal lobe attacks (a notion the other specialist discounted). The family had seen little point in registering with a local doctor in London before referring the children to another specialist. It took a shrewd paediatrician with an interest in family dynamics to recognize the profound marital problem which existed. Neither child was found to have any serious underlying physical problem, although both showed signs of considerable emotional distress. In effect, their symptoms signalled their parent's distress which Mr and Mrs Conrad had found painful to acknowledge. Help was offered in the form of family sessions from a therapist from which all parties obtained immediate benefit.

From this one case it must not be assumed that children's problems are always extensions of parental and marital difficulties. Children can *create* problems both for themselves and for the family. Nor is it true to say that where there is a family problem, every family member needs treatment. Here, focusing on the marriage and on the parents' difficulty in adopting a consistent approach to parenting did improve the children's behaviour. In another case, helping to resolve relationship difficulties between parent and child led the parents themselves to focus on their own relationship difficulties. In a third situation, where mother and father lived apart and could not come to an agreement about parenting their only child who was depressed, an improvement was obtained by focusing individually on each parent in an attempt to boost self-esteem.

RECOGNIZING EXPLOITATION OF CHILDREN

Exploitation refers to an extreme situation where one person takes advantage of another because of differences in age, power, or authority. It is easy to see how children throughout the ages have been exploited by their parents. The recognition of what amounts to serious family disturbance can range from obvious physical injuries to more subtle situations such as emotional neglect and sibling incest.

Child sexual abuse is the ultimate abuse of adult power. On more than 60 per cent of occasions the perpetrator is a close family member. The abuse needs to be attended to for both the child's sake and for the sake of the families which fail to protect their children. Both need help. General practitioners should never attempt to tackle their problem on their own, and each area of the country has procedures laid down for referral and management of such cases. Collaboration between disciplines is frequently fraught—a reflection of the splitting that often goes on in the families of abused victims. Nevertheless, a multi-disciplinary approach backed up by statutory legal measures provides the best means of helping such families and the victims of abuse.

The escalation in the numbers of cases reported and the realization that the morbidity associated with child sexual abuse is very great has led to interest in ways of improving recognition. General practitioners are often the first professionals to suspect that abuse has occurred. In the surgery concern can be raised in two ways:

1 By the parent (usually the other) raising the issue with the GP.
2 As a result of examination of the child.

Leventhal *et al.* (1987) have outlined ways in which the possible diagnosis of sexual abuse can be raised with parent and child, and points out the importance of attaching weight to what the child, in particular, is saying. In their opinion fictitious allegations of abuse are rarely made by children. The family needs to be handled with sensitivity. The manner in which the subject is approached, the person who is being questioned, and the context in which the enquiry is being conducted all have a bearing on the outcome. Having a straightforward, open-minded approach is essential. Using the examination of the child both as a diagnostic tool and as a means of enabling the child to speak about his fears and anxieties can be helpful. Tact is needed, too, with the parents, especially if one of them is suspected of being the perpetrator. Occasionally it is with a sense of relief that a family discloses its secrets to the professional, but more commonly one is faced with denial.

It is to be hoped that by creating an atmosphere of trust and concern in which feelings of guilt, fear, and helplessness can be expressed, families with problems as serious as those which involve the exploitation of children will come for help earlier than they do.

7 Understanding families

If we are properly to understand and work with family problems, we need a frame of reference to orientate ourselves.

Dare (1979) suggests thinking about families in three basic ways:

1 Seeing the family as an entity which is constantly evolving and developing; in other words, paying attention to the 'family life cycle'.
2 Recognizing the influence of successive generations on each other and noting the way patterns of behaviour in families can become repeated. This involves seeing the family in an historical context.
3 Regarding the family as a 'system', or a complex of elements in mutual interaction with one another.

Let us consider each in turn.

FAMILY LIFE CYCLE

The concept of a family life cycle rests on two observations:

1 That families experience changes through time in similar and consistent ways.
2 That variations in the degree to which one family member is dependent on another are related to the stage which that family has reached in its life cycle.

Duvall (1977) distinguishes eight stages in a family life cycle (Fig. 7.1).The transition between one stage and the next is a potential time of difficulty both for the individual and for the family as a whole. It may involve anxiety, but whether or not problems arise, it will invariably be a time of adjustment for several family members (see Table 7.1).

Take, for example, the changes involved for a family whose child is entering adolescence. The whole family has to accommodate to the changes that the adolescent has to achieve through desire for independence, discovery of new relationships, and awareness of adult responsibilities. Any difficulties created by the adolescent are by definition family difficulties, resulting in conflict between family members as well as conflicts within the adolescent himself or herself.

Within each stage of the family life cycle specific 'tasks' common to each individual can be recognized. Such 'tasks' need to be accomplished if problems are not to arise later on. Thus a couple at Duvall's stage 1, who are beginning a family, need to work towards achieving the following:

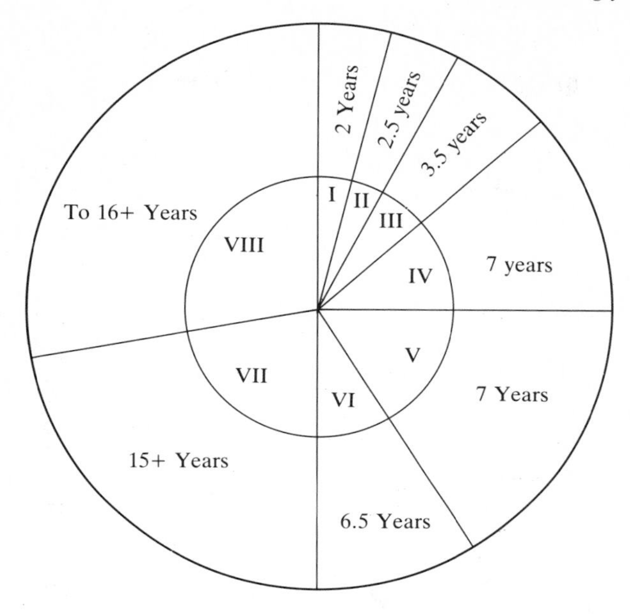

Phase	Family phase	Family description
I	Beginning family	Married couple without children
II	Childbearing family	Oldest child, up to 30 months
III	Families of pre-school children	Oldest child, 30 months to 6 years
IV	Families with school children	Oldest child, 6–13 years
V	Families with teenagers	Oldest child, 13–20 years
VI	Families as launching centres	First child gone to last child leaving home
VII	Families in the middle years	Empty nest to retirement
VIII	Ageing families	Retirement to death of both spouses

Fig. 7.1 The family life cycle. Glick, I. D. and Keffler, D. R. (1980). *Marital and family therapy*. 2nd Ed. Grune and Stratton, Inc., p. 36. *By permission of the Psychological Corporation, Orlando. Florida*

- A home;
- An income;
- An agreement as to who does what and who is accountable to whom;
- A satisfactory sex life and system of communication;
- A workable relationship with other family members;
- Some system of working satisfactorily with friends and colleagues outside the home;
- A plan for the future which takes children into account;
- Some workable philosophy of life.

Duvall's life cycle applies to most families, though matters have become more complicated as new types of family structure have emerged. In the

Table 7.1 *Developmental tasks through the family life cycle*

Stage of the family life cycle	Positions in the family	Stage-critical family developmental tasks
1 Married couple	Wife Husband	Establishing a mutually satisfying marriage. Adjusting to pregnancy and the promise of parenthood. Fitting into the kin network.
2 Childbearing	Wife–mother Husband–father Infant daughter or son of both	Having, adjusting to, and encouraging the development of infants. Establishing a satisfying home for both parents and infant(s).
3 Pre-school age	Wife–mother Husband–father Daughter–sister Son–brother	Adapting to the critical needs and interests of pre-school children in stimulating growth, growth-promoting ways. Coping, with energy depletion and lack of privacy as parents.
4 School age	Wife-mother Husband–father Daughter–sister Son–brother	Fitting into the community of school-age families in constructive ways. Encouraging children's educational achievement.
5 Teenage	Wife–mother Husband–father Daughter–sister Son–brother	Balancing freedom with responsibility as teenagers mature and emancipate themselves. Establishing postparental interests and careers as growing parents.
6 Launching centre	Wife–mother–grandmother Husband–father–grandfather Daughter–sister–aunt Son–brother–uncle	Releasing young adults into work, military service, college, marriage etc., with appropriate rituals and assistance. Maintaining a supportive home base.
7 Middle-aged parents	Wife–mother–grandmother Husband–father–grandfather	Rebuilding the marriage realtionship. Maintaining kin ties with older and younger generations.
8 Ageing family members	Widow/widower Wife–mother–grandmother Husband–father–grandfather	Coping with bereavement and living alone. Closing the family home or adapting it to ageing. Adjusting to retirement.

Source: Duvall, E. M. (1962). *Marriage and family development*, 5th Ed. Reprinted by permission of Harper and Row, Publishers Inc. New York

case of reconstituted families, for example, the 'tasks' are more numerous and require a lengthy and difficult series of adjustments for all members.

An awareness of a family life cycle can be a useful predictor of behaviour arising out of alterations in family equilibrium.

Alongside the ordinary 'passages' in life, such as the move into adolescence, unexpected losses have to be contended with. These can further disrupt the family and pose sets of difficulties for each member. The loss of a close family member constitutes a particularly important time of transition for those who are left, but not every loss is harmful. Doctors are aware of people who weather bereavement and emerge stronger and more mature than they were before. Moreover, some losses pass without the expected disturbance to the close relative who is left. However, as we saw in a previous chapter, loss can endanger both mental and physical health.

Clearly the more unexpected losses are encountered in addition to the critical transitions in the life cycle, the more likely it is that the family as a whole will experience disruption and that individuals will need to consult their doctor. In the chapter devoted to working with family problems, emphasis is placed on actively looking for the changes in family relationships which are brought about by painful transitions.

THE INFLUENCE OF SUCCESSIVE GENERATIONS

In psycho-analysis much emphasis is placed on the earliest family influence of all, the relationship between mother and child. Personal development, however, is the product of experiences from all phases of life. These become internalized and affect a person's expectations in relationships later on. In addition, the influence which would form a person's life can stretch back over many generations and may take the form of family 'myths' or 'legends'.

Pincus and Dare (1978) talk of a family 'myth' as being 'a secret or unconscious belief, or attitude which, through its general acceptance by succeeding generations of a family, comes to perpetuate itself in determining responses and behaviour'. Family 'legends', on the other hand, are stories which convey to each generation the rules and obligations of that family. They have a moral signficance and, though each telling produces a version different from the one before, the moral message in them remains the same.

Some family therapists use family 'myths' and 'legends' as tools for change.

Growing up in a family provides us with models of partnership and parenting which remain for the duration of our lives. They play a critical role in organizing our later relationships as partners and parents. For example, one of the determinants of marital choice is our own internal representation of the family in which we are brought up. Even a deceased relative can act as a force in marital choice, should feelings once reserved for that partner be re-evoked in a new relationship. Some individuals select partners whose traits are clearly similar to those of their parents; others are led in an altogether different direction and appear to select partners whose attributes contradict the parental image. Robin Skynner and John Cleese (1983) discuss the marriage of 'opposites' in their entertaining conversation pieces, *Families and how to survive them*:

Robin: I think the reason that we're attracted to someone at a deep level is that basically they are like us—in a psychological sense.
John: But the wise saw tells us 'opposites attract'.
Robin: They don't, or if they do, its because they seem to be opposites. But really what draws people together is their similarities, and moreover a similarity in one of the most fundamental aspects of all—that of their family backgrounds.
John: You mean all these people who used to get married—and still do—to escape from their families are in some way taking their families with them, psychologically speaking?
Robin: Exactly.

Many marital problems can be viewed as family problems which have been emotionally and psychologically inherited and perpetuated, a theme which is elaborated in Chapter 11.

A SYSTEMS APPROACH TO THE FAMILY

Thinking along systems lines can be used to understand better some of the processes that go on in families: the interaction of family members with each other; the levels of organization tht exist within families; and the sense of 'wholeness' that a family conveys.

Von Bertalanffy (1968), a biologist, first developed general systems theory in the 1930s to account for some of the phenomena he saw in the physical as well as the biological world. A 'system' he defined as 'a complex of elements in interaction with one another', and he recognized two types, 'open' and 'closed'. 'Open' meant open to influences from outside. The family is a good example.

Families are affected from without by friends and relatives as well as by social and political systems. They, in turn, can exert an influence on the outside world. Consider the profound effect a tragic death can have on neighbours, friends, and even the community as a whole.

Each family is however a discrete entity. It has a **boundary**—not a physical one, but an emotional one which marks if off from other families. What is more, within each family system can be recognized **sub-systems**, each protected by its own boundary and having an identity of its own.

Hence we have husband–wife, parent–child, brother–sister, male–female sub-sets (Fig. 7.2).

A boundary serves to restrict emotional interchange between what is within and what lies outside. So called structural family therapists concern themselves particularly with the types of boundaries that can exist between family members. Two extreme types—'enmeshed' and 'disengaged'—are recognized (Fig. 7.3). In an enmeshed relationship the boundaries are blurred and the behaviour of one family member has an instant and marked effect on others. In a disengaged relationship, the boundaries are more clearly drawn and the behaviour of one family member has little effect on another. Most families fall somewhere between these two extremes.

Within a single family, it is possible to have both enmeshed and disengaged relationships side by side. For example, a mother–child sub-system often displays enmeshment when the child is young with the father assuming a disengaged position with regard to the child.

In their extreme form, enmeshment can undermine independence while disengagement can result in isolation. Minuchin (1974) puts it another way: 'The parents in an enmeshed family may become tremendously upset because a child does not eat his dessert, while parents in a disengaged family may feel unconcerned about a child's hatred of school.'

Despite the fact that boundaries allow different inputs to affect the system within, there is a tendency within all open systems to maintain a 'status quo'. This phenomenon, known as 'equifinality', is familiar to students of the human body who learn about the regulatory mechanisms which keep the body biochemistry the same despite a constant flow of material in and out of it. In families similarly there is a tendency to accommodate to disturbance. Despite this, change is possible and because its effects can reverberate around the entire family, it provides therapists with the opportunity to bring about shifts in behaviour, not just in one person but in many.

An important property of an open system is that its parts are *inter-dependent*. For this reason a family is more than the sum of the individual personalities who comprise the household; it includes the relationships between the members, and has a 'wholeness' of its own.

Systems thinking above all helps us escape from the view that all problems are the property of the individual who presents them. More often than not they belong to the larger system, the family.

Whatever basic ways of thinking we use in understanding families—and there are other ways based on communication and power—it is always necessary to take account of the social setting. Not only will it differ with every family but the degree to which it influences that family will vary. 'Looking at the family apart from its social context', as Minuchin (1984)

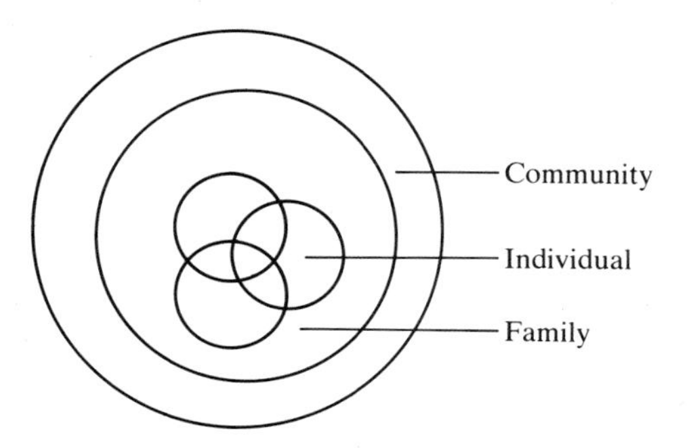

Fig. 7.2 Systems within systems

Disengaged position
Clear boundaries separate all three

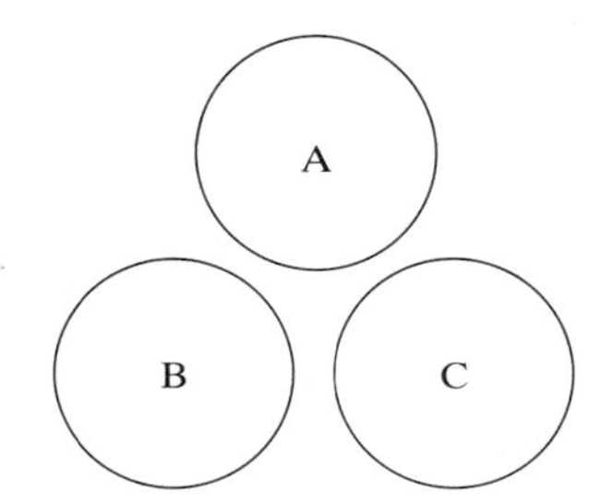

Boundaries between A and B, and A and C are blurred.
That between B and C is clear

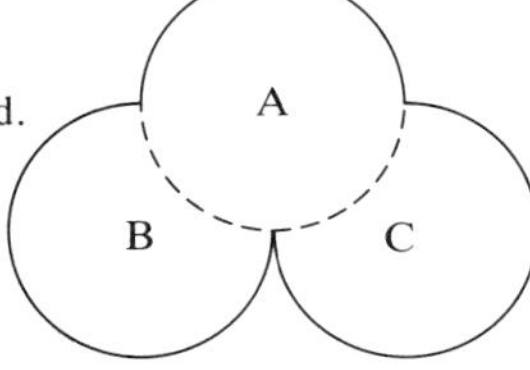

Boundaries between all three are blurred — enmeshment.

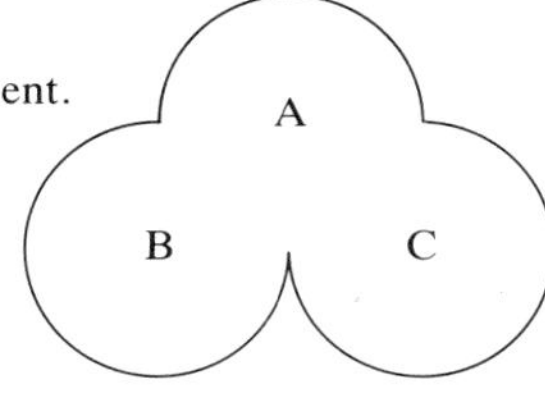

Fig. 7.3 Boundaries

says, 'is like studying the dynamics of swimming by examining a fish in a frying pan'. Here, the work of the family doctor comes into its own, since knowledge of the environment and local community can help understand the behaviours seen in the surgery and in the patient's home. If there was ever a reason to justify home visiting, it is this.

8 Working with family problems

When confronted by the pain and anguish some families experience, theories as to the cause of their problems appear to offer little guidance when it comes to intervention. Sharing a problem can minimize it but merely listening to the patient's distress often seems inadequate. Offering a prescription may seem the only other solution; yet instinctively we know it solves little, offering no more than a token of our concern and a way of bringing the consultation to an end. Equally, if we are to be too concerned and helpful, there is a risk that the individual comes to rely on us for solutions to his problems and is prevented from discovering for himself his potential for change.

CHANGE CAN BE PAINFUL

By intervening, we are interested in producing change and change for the better. Patients wish for change, too, if only to relieve themselves from distressing and disabling symptoms. But most change is painful, and what confronts the doctor and makes him feel frustrated is the dilemma that, despite people's wish to change, they often experience a contrary desire to keep things as they are. This may reflect a resistance on the part of others to alter the status quo. In systems terminology, the family operates a self-correcting mechanism which seeks to maintain a state of equilibrium so as to prevent wholesale disintegration.

In promoting change we appeal to people's willingness and capacity to grow and develop. It is often easier to achieve lasting change by involving the whole family than by working with one individual, the presenting patient. The reason is that it is the family which is at fault rather than the individual who holds the symptoms, and real improvement relies on changing the family system as much as relieving the patient of his symptoms. Some families will, of course, cling to existing behaviours as being the best solution to their problems and keep things as they are whatever the cost, but many others will be prepared to countenance change if provided with the necessary information and understanding which allows them to overcome difficulties. It is the therapist's task to try to provide what is missing.

FIRST MOVES

Convening the family is something to aim at within one or two consultations, although there may be problems in engaging all family members.

In the lead up to a family interview, it is often useful to construct a 'family wheel' (Williams, 1982). The presenting patient is invited to comment on each member of the family in the household. Remarks which reflect on the quality of the relationship with others are recorded verbatim, for example 'A pall of silence falls over us when father enters the room', or 'He never seems to be there when it matters'. Such remarks often succinctly sum up problems that exist in the household. Although simple, they often preserve an underlying complexity of ideas and serve as comments that can be taken up at a later point. Patients are often impressed when a doctor quotes something from their own lips at some later stage, especially in general practice, when several consultations are involved, and there is a need to demonstrate continuing concern and to memorize what has gone before.

The presenting complaint is placed in the inner circle and remarks made about each family member recorded against their place in the outer circle. Sometimes it is useful to draw lines of support (straight) and of tension (wavy) between individuals. The home as a unifying or disruptive influence on the health of its members is accorded space, too, since family atmosphere as well as social circumstances may play a part (Fig. 8.1).

In the case of children who present with management or behavioural difficulties, the model can be used in a similar way (Fig. 8.2), but here the child occupies the central space and remarks made about the child by the parents and siblings are recorded against their names. Again lines can be drawn between members, and comments which reflect the nature of the relationship between two partners written adjacent to these lines. Since emotional climate of the home is often summed up in a remark made by one of its occupants, space is again given for this in the wheel. It can also be a useful exercise to write down a plan of action, even if this is revised in the light of seeing the whole family together, since it encourages the doctor to think about the aims of treatment.

A device like a 'family wheel' assists general practitioners in thinking more clearly about the problems presented to them, and is a form of classification which brings together important elements:

- the presenting symptoms;
- the emotional climate of the family;
- attitudes towards the presenting patient reflected in off-the-cuff remarks;
- the quality of the relationship between near family members with points of tension and harmony brought into focus.

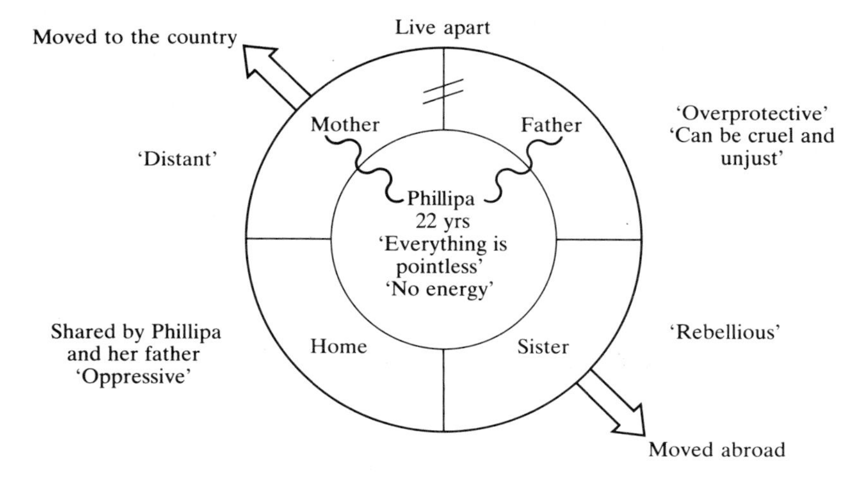

— Define parental influence which causes Phillipa to stay and her sister to leave.
— Assess Phillipa's mental state and her willingness to accept help

Fig. 8.1 Family wheel. Example 1.

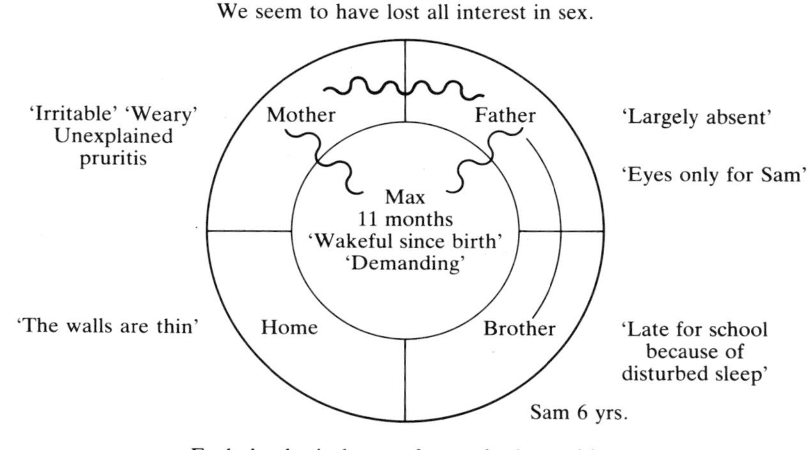

— Exclude physical cause for mother's pruritis.
while remembering other 'irritants' in the family
— Allow male jealousy to be expressed.
— Create time for the couple's joint needs.

Fig. 8.2 Family wheel. Example 2.

Stamps of 'family wheels' can be made which make an inked impression in the notes over which the comments can be written with a pen. The 'family wheel' is a first step in shifting the perspective from the individual's presenting symptoms to relationships within the family and leads naturally on to the family interview.

CONVENING THE FAMILY

This requires careful planning. An important preliminary is to offer reassurance that the object of involving close family members is not to blame them but to seek their help in solving what has become a shared problem. Some may doubt the notion that problems can arise out of a disturbed family interaction—so to leap in with suggestions that others are involved can be a strong disincentive to further attendance. Tact and sensitive handling are needed, and it may be some while before a readiness is shown to involve others in what is seen as a personal problem. Convening is an important step forward and there are several reasons why as many family members as possible should attend:

1 The problems of one individual are normally best understood when placed in a family context.
2 The problem is often a shared one.
3 The family itself may hold the key to successful solutions.

First contact should be non-threatening and designed to establish a sympathetic relationship between doctor and family. This process has been termed 'joining' the family and a good way to start is by constructing a 'genogram' or family tree.

The genogram

Some notion of family structure and relationships will have been obtained from the 'family wheel'. The 'genogram' goes one stage further. It records information about all family members and their relationships over at least three generations. As such it provides a summary of the family history. It is constructed jointly with the whole family and added to or modified on subsequent encounters. It helps all parties to think in a systematic way about their problems and demonstrates—often forcibly—repetitive patterns of illness and behaviour. Families do repeat themselves. Similar issues tend to be grappled with by different generations. Genograms bring together the current and the historical context of the family and enable families to see their problems as not mere random events but as connected with other difficulties. It quickly becomes apparent how family problems tend to cluster around transitional stages in the family life cycle.

Each family is represented by a box or circle according to sex, with the identified patient accorded a double line. People who are dead have a cross placed inside the symbol, and date of birth and death are shown to the left and right above each figure. Pregnancy, stillbirth, miscarriage, and abortion are given special symbols, with lines drawn between individuals signifying ties. Convention has it that the husband is on the left and the wife on the right, joined by a line which has above it the year of the marriage. Separation and divorce are shown by a line or lines cutting across the marriage line, again with a date inserted. Unmarried couples are joined by a dotted rather than a continuous line. Children belonging to a couple connect to the line that joins the two adults with the eldest child on the left and the youngest on the right. Members who live together under one roof may be encircled by more dotted lines (see Table 8.1). Some examples of genograms are shown in the accompanying figures (8.3, 8.4, 8.5).

Having drawn the family structure in this way other information is added—any that is considered important, for example, occupation, place of living, signficant life events, illnesses, or disabling conditions, etc.

Gathering such information during an initial family interview should involve everyone present. They may correct each other, supply information not known hitherto, and provide important observations on the way their family works. It is the doctor's task to bring this about, and also to focus attention on the current state of affairs, so that this is not lost in the welter of information supplied about previous generations. It is his job, too, to ask more searching questions about relationships—who gets on well with whom, who are in conflict, what members are particularly close to one another, who does the caring, and who rules the roost. Once a relationship has been established between doctor and family, it may be possible to enquire into sensitive areas such as drink problems, problems with the law, and emotional illness. The information supplied can then be examined for:

1 Repetitive patterns;
2 Any coincidences, for example, the anniversary of a loss coinciding with an onset of symptoms in another individual;
3 Any change in relationship brought about by critical life events or transitions in the life cycle.

Tomson, a practitioner with considerable experience of using genograms in general practice, stresses the importance of the manner in which the session is conducted (Tomson 1985). Sharing in the enthusiasm with which members map out their family tree, and even offering vignettes from one's own life helps. A flip chart, or a piece of foolscap, is used to plot the genogram, and this is then handed at the end of the session to the family, making it clear that it is their property rather than the doctor's. Those present are then able to share the record with any family member who did not attend. Only with their permission does Tomson make a copy for himself. He appears involved and yet non-judgmental, and effectively uses

Table 8.1 *A table of symbols used in constructing genograms*

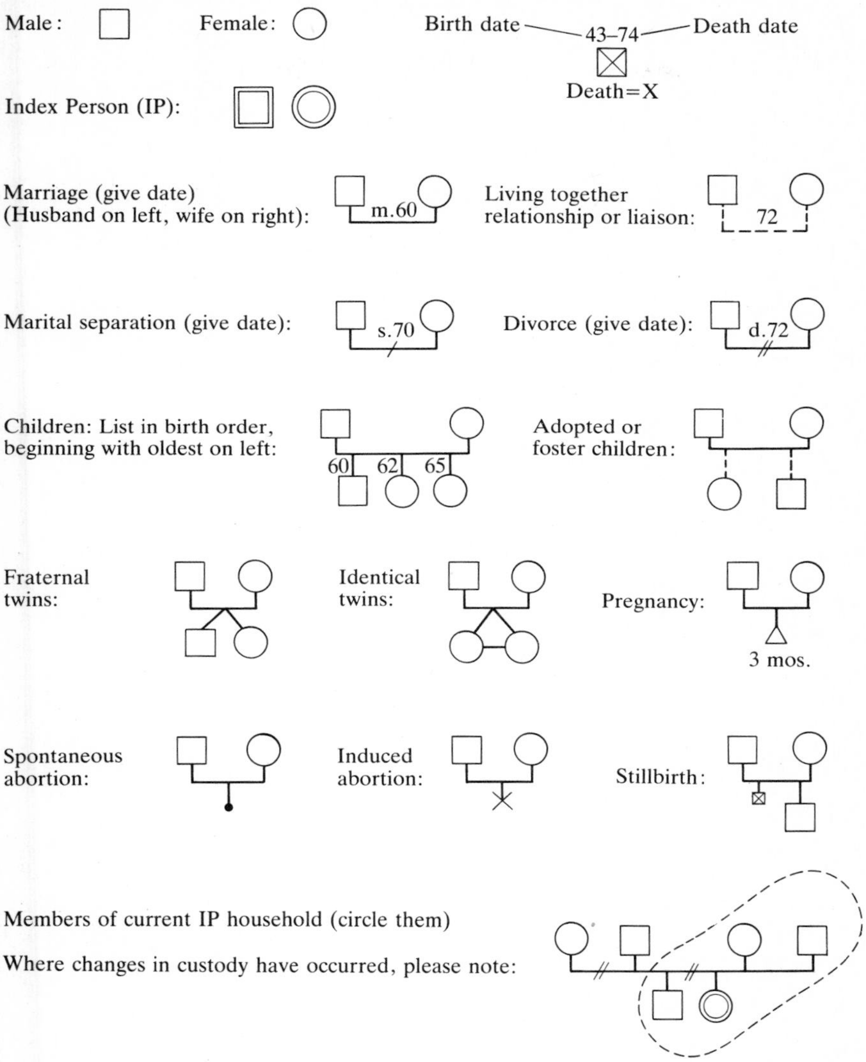

Male: □ Female: ○

Birth date — 43–74 — Death date

Death=X

Index Person (IP):

Marriage (give date)
(Husband on left, wife on right): m.60

Living together
relationship or liaison: 72

Marital separation (give date): s.70

Divorce (give date): d.72

Children: List in birth order,
beginning with oldest on left: 60 62 65

Adopted or
foster children:

Fraternal
twins:

Identical
twins:

Pregnancy: 3 mos.

Spontaneous
abortion:

Induced
abortion:

Stillbirth:

Members of current IP household (circle them)

Where changes in custody have occurred, please note:

Reprinted from *Genograms in Family Assessment* by Monica McGoldrick and Randy Gerson, by permission of W. W. Norton & Company, Inc. copyright © 1985 by Monica McGoldrick and Randy Gerson.

the exercise as a tool for 'joining' the family. The extent to which this is successful is in the way which people often are prepared to reveal 'secrets' about themselves—previous abortion, undeclared relationship problems, alcohol problems, etc. The act of sharing assists them in coming to terms

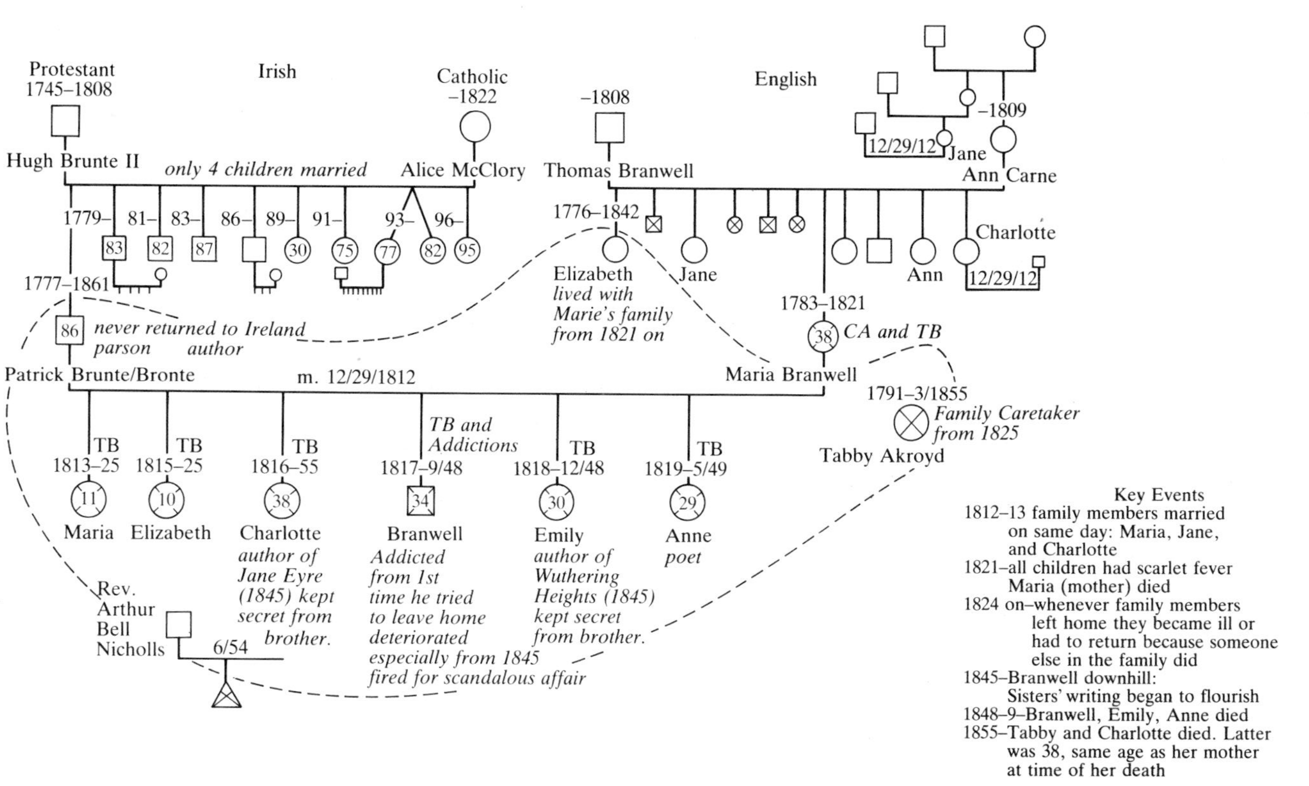

Fig. 8.3. Genogram of the Bronte family. The genogram shows how all six Bronte children were born over a seven-year period. Their mother died two years after the death of Anne, the youngest, and they were reared in virtual isolation from other children. Three became accomplished writers. All experienced difficulty leaving home. Only Charlotte married, at the age of thirty-eight, and she died after only nine months of marriage. Whenever the children tried to leave home they became symptomatic and returned. For the Bronte children leaving home and establishing families for themselves was the problem. (Reprinted from *Genograms in Family Assessment* by Monica McGoldrick and Randy Gerson, by permission of W. W. Norton & Company, Inc. copyright © 1985 by Monica McGoldrick and Randy Gerson.)

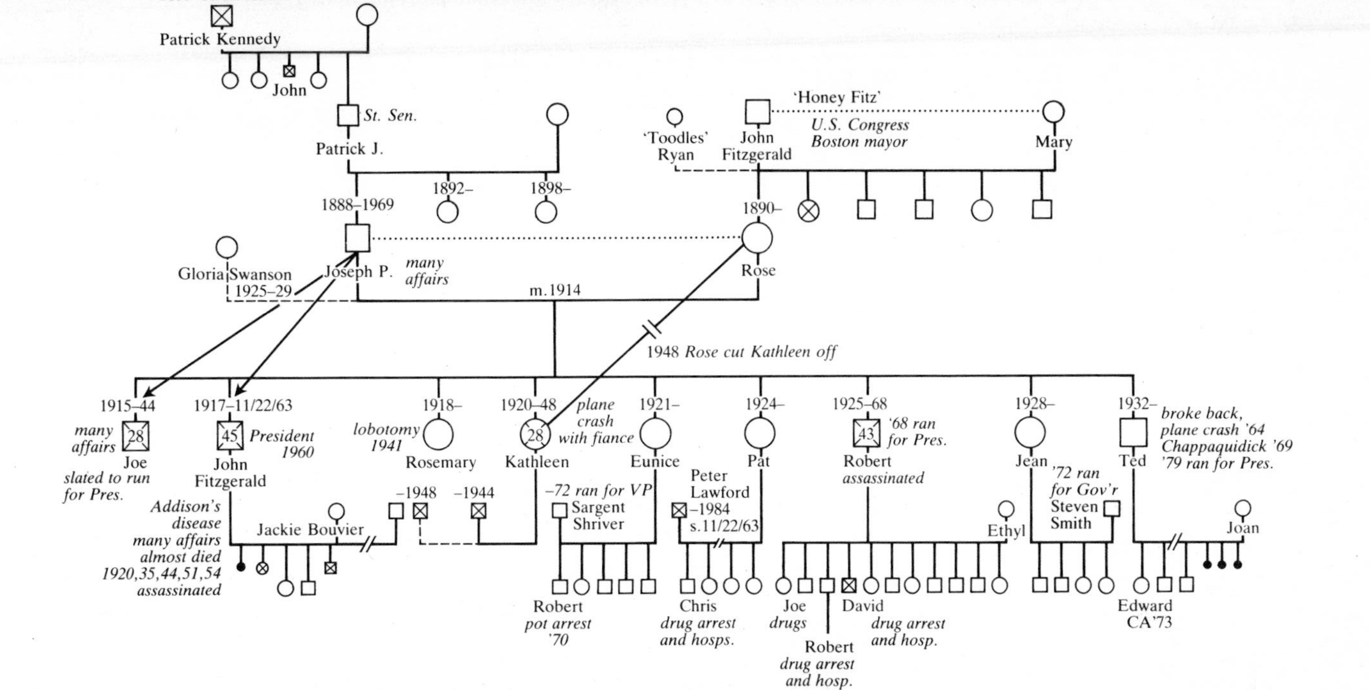

Fig. 8.4. Genogram of the Kennedy family. The Kennedy family genogram is remarkable for the extraordinary number of premature deaths and personal tragedies. To what extent events related one to another is debatable, but it is certainly the case that stressful life events increase susceptibility to accidents. Thus, Ted broke his back in a plane crash seven months after John was shot, and was involved in the Chappaquidick incident in which one person drowned twelve months after Robert was killed. The genogram illustrates the way critical life events send ripples through the family system. Of the 29 grandchildren, one died of an overdose of drugs, and at least four others have had drug arrests and/or psychiatric hospital admissions. These include four of the six eldest sons suggesting the pressures on the oldest son are greatest in such families. (Reprinted from *Genograms in Family Assessment* by Monica McGoldrick and Randy Gerson, by permission of W. W. Norton & Company, Inc. copyright © 1985 by Monica McGoldrick and Randy Gerson.)

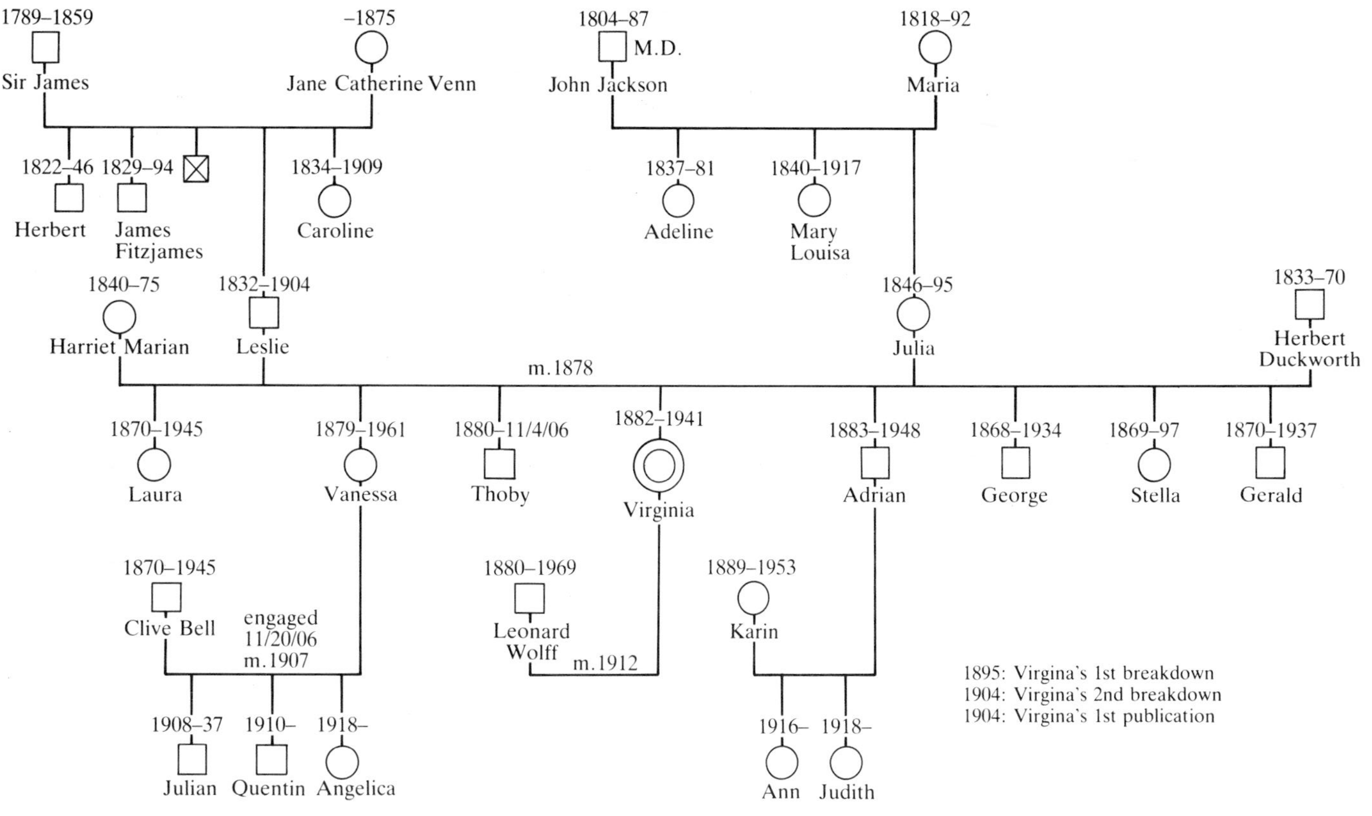

Fig. 8.5. Genogram of Virginia Woolf's family. This illustrates how events in a family can be interconnected. Virginia's two mental breakdowns occurred shortly after the deaths of her mother and father, and Vanessa became engaged two weeks after her brother's death. (Reprinted from *Genograms in Family Assessment* by Monica McGoldrick and Randy Gerson, by permission of W. W. Norton & Company, Inc. copyright © 1985 by Monica McGoldrick and Randy Gerson.)

with their feelings and in recognizing the influence of others close to them on their behaviour. The exercise also sows in people's minds the seeds of change. It offers them the potential for altering their behaviour since patterns which have been learned and perpetuated by generations are no longer seen as being fixed but are capable of alteration. As Tomson puts it, it can 'stimulate thoughts about relationships and give permission for changes to happen'.

Clarifying objectives

In subsequent sessions, either with the whole family or a selected group, it is important for the doctor to clarify his objectives. First he must decide to what degree he is prepared to work with the family, and if not, whether he wishes to refer the case, and if so to whom. His objective may simply be to make an assessment of the problem in the light of the family dynamics and to assess people's motivation in seeking help. He may go further and ask individuals directly what they want to see made different, and may even present them with a series of alternatives from which to chose. His next task is to match the needs of the patient, family, or couple with any appropriate form of help. Because there exists no uniform theory of family disorder, the doctor has a wide range of techniques and therapies from which to chose. Some he may accomplish himself with a modicum of training and experience, others are rightly the province of the specialist.

9 Working with entire families

Working with families, as opposed to working with individuals, involves the doctor in new ways of relating to patients. His aim remains the same—to bring about change where possible.

For most family therapists behavioural change is the prime goal. They are less concerned with imparting insight than with encouraging new behaviours to replace those that are damaging to the family as a whole. This can be true also of the general practitioner who works with family groups, but achieving a shared understanding of a problem and supplying insight into the way family members react towards one another can, of its own, produce subtle behavioural shifts. It is fortunate that there exist families who are prepared to change in limited ways with the help of relatively inexperienced practitioners, and it is for this group of doctors and patients that this chapter is primarily written.

THE ROLE OF THE GENERAL PRACTITIONER

The general practitioner starts with certain advantages:

1 His intimate knowledge of family members in a wide variety of settings—at home, in the surgery, seen together and separately, when ill and when in rude health. This provides important clues not only about individual psychopathology, but also about the relationships of family members at times of crisis and relative calm.
2 His presence at a birth, or on the occasion of a bereavement, or merely as a witness to a conversation over an ill relative gives him a wide understanding of the workings of many of the families on his list.
3 If he works in a team he is likely to gain from the insights of other members of his team and build up a composite view of families who otherwise are only glimpsed through the actions and remarks of individual members.

The general practitioner also labours under certain potential difficulties:

1 Many families split themselves between more than one doctor, so that the general practitioner may not 'own' more than one or two in a family group, and he needs the permission of other doctors as well as their patients before working with whole families.
2 Doctoring the mind as well as the body is something that should come easily to most general practitioners, but being the gynaecologist to one

member of a family, physician to another, and 'father-confessor' to a third, can create difficulties for him when he becomes therapist to them collectively. Issues of confidentiality can confuse the situation sufficiently for him not to want to see all members together.

Leaving aside these potential difficulties, as long as the G.P. sets himself reasonable goals, he can still help considerable numbers of families who experience difficulties in communication, problems with parenting, and trouble in adjusting to changes in their circumstances provoked by loss and illness.

In working with entire families, a **common goal** is essential. If achieved, conflict is likely to be reduced and those bearing the brunt of the problem will feel released from a burden. But the doctor must be careful not to impose his own standards on families who have a quite different set of values and requirements from his own, and, as Skynner points out, he must be prepared to question to himself, not just the likelihood of achieving change, but the desirability of change itself in any family situation.

A METHOD

I shall assume that the general practitioner has a room at his disposal large enough to accommodate the immediate family, and is able to set aside forty minutes of his uninterrupted time for each of six sessions.

Using the three basic ways of thinking discussed in Chapter 6,* he will already have arrived at some understanding of the family he is dealing with and now needs to define with each person in turn what is seen as being the main problem and in what way they wish things to be different, individually and collectively.

To understand the conflicts that exist between members, he must become part of the family system for a time and experience directly for himself the demands that the family make on him. He should show a readiness to work with everyone present and try and understand the problems from each person's point of view. This is not as daunting as it sounds and each doctor should be allowed to develop his own individual style through experiment.

Simple observations of the way individuals behave towards one another—where they sit, who speaks to whom, who interrupts, who supplies most information, and who remains silent—can all be of assistance. There is often an 'appropriate distance' between individuals which

* Three basic ways of thinking about families include paying attention to the 'family life cycle', recognizing the influence of successive generations and regarding the family as a 'system'.

It is assumed that the GP will have already arrived at some measure of understanding using the three frames of reference discussed in Chapter 6, and will have constructed a 'family wheel' or a 'genogram' with one or more members of the family.

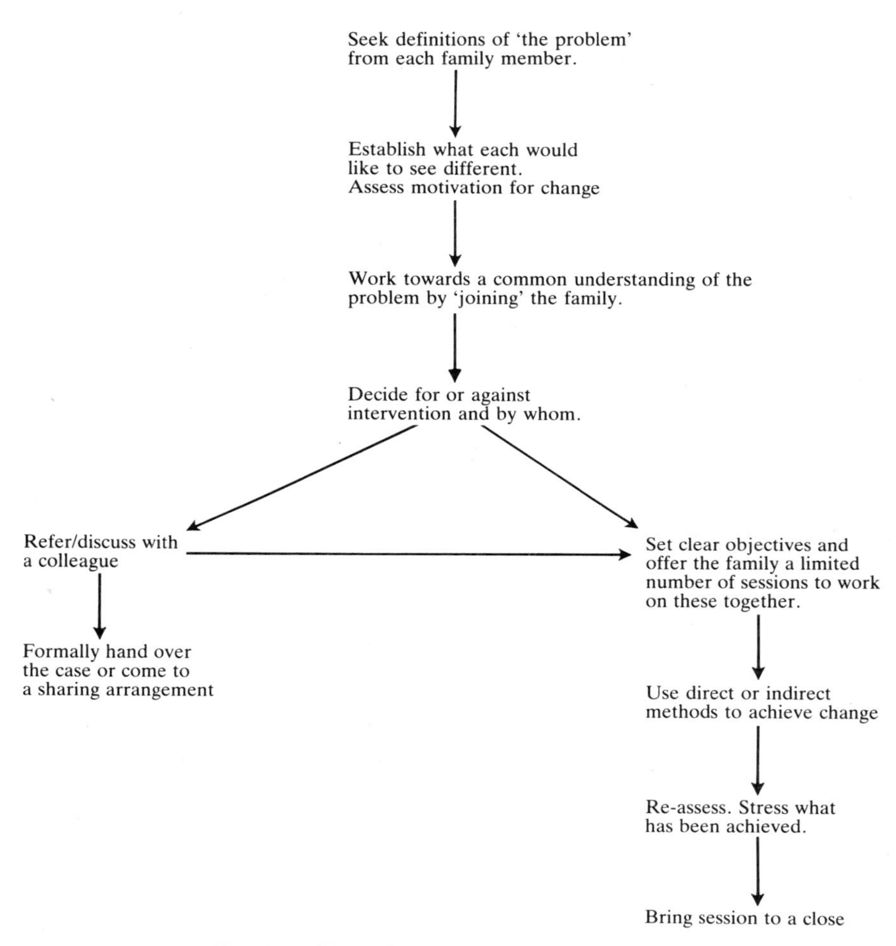

Fig. 9.1 Plan of working with entire families

is reflected in the way people arrange themselves in the room; physical distance reflecting psychological distance.

Usually, one member is identified early on as having a particular problem. This should not blind the doctor to focusing on others and a useful way of diverting attention away from the 'sick' individual is to enquire in what way his behaviour affects them individually and their behaviour towards one another. It is because progress can be inhibited by having too sharp a focus on one person that the doctor has to take an active control over the consultation, so preventing one person or pair from dominating the session. There will be occasions in which individuals are kept somewhat peripheral in the family discussion and require the support of the doctor to speak out and challenge what the others are saying. This can pose problems for the doctor, who on the one hand seeks to dispense favours equally and be non-judgemental, but on the other, challenge the family and even champion the cause of an oppressed member. By not intervening he risks preserving the existing family symmetry and limiting alternative views, while by intevening he stands accused of showing favouritism to one person. Since the essence of treatment is often to challenge a family's view of the problem, he may opt for intervention.

The method he uses needs careful thought. A technique such as 'unbalancing' where the goal is to change the power structure of the family through personal involvement, for example by forming a coalition against one or more family members, is largely unsuited to general practice. Another exercise, which highlights boundaries between people, such as physically placing an 'outsider' in the family at a distance from the others in the room, can be a clear and direct way of making a point without using words and cause the family to rethink their emotional closeness to one another. Most people understand the metaphor of proximity as an expression of emotional closeness; changing the seating arrangements in the room to emphasize these divisions is a very practical method to employ in a room of reasonable size. Moreover, such a method is applicable to all social groups, for example, those middle-class families who are apt to intellectualize their problems can benefit greatly from such an experimental technique.

Suggesting to people that they behave in certain ways—for example, telling parents to set limits to their child's behaviour and co-operate in seeing that these are obeyed—can be effective. Often, however, direct approaches fail and indirect methods need to be used. The use of metaphor is itself an indirect method of communicating ideas, but there are two other 'strategic methods' which lend themselves particularly to use in general practice.

Paradoxical methods

Telling someone to do the opposite of what might seem logical can sometimes have surprisingly beneficial effects. Take the example of 'prescribing the symptom'. The doctor suggests to an over-protective parent that she actually increase her attention towards her child, and to the wife who feels she has to watch over her husband in case he misbehaves, that she intensify her behaviour towards him. Should these injunctions be obeyed, the likelihood of conflict between the participants is increased but this is usually followed by a distancing effect—the desired goal.

Paradoxical ways of working include:

- Giving the family both permission to carry on behaving in a dysfunctional way and an injunction to intensify their symptomatic behaviour to the point of crisis.
- Predicting that disputes will recur on particular occasions in the future.
- Taking the problem more seriously than the family does itself.

Such methods are particularly applicable in situations where the family is stuck for a solution. The repetitive patterns of behaviour seen in some families prevent other more adaptive behaviours from emerging, and paradox can enable certain families to break out of the constraints that deny change.

Reframing

Reframing a problem in a positive light is a way of altering a family's perception of a symptom or behaviour. Thus anger on the part of an individual is interpreted as 'caring behaviour', a demonstration of 'concern'; while a particular behaviour which keeps a symptom alive in the family is redefined as a force which keeps a family together and stops it from fragmenting.

To become an effective agent of change involves the doctor in learning some of the skills of family therapists. Handbooks such as Minuchin and Fishman's *Family therapy techniques* (1981) can be useful, but there is no substitute for experience and being prepared to try out different approaches. Each of us needs to find a 'comfortable' method of operating and yet be alive to alternatives.

ENDING

It is not usually necessary or indeed desirable to work beyond six sessions with any family, and these should ideally be spaced by two-week intervals to allow for developments. In terminating it is important to take stock of such shifts as have occurred, giving praise to the family for achievements, but also highlighting those areas which remain unresolved. It is not intended that any final solution be achieved but rather a break-through in communication and understanding which can subsequently be built upon by the family itself. Hopefully the family, as well as the doctor, will have achieved, at a bare minimum, a clearer understanding of the problem they share and be prepared to discuss matters more openly and directly when things go wrong.

THE FURTHER WORK OF FAMILY THERAPISTS

It is not my purpose here to describe the workings of various family therapists from whom we can learn and to whom we may refer cases. To begin with, the choice is usually limited to what is locally available, and that confined to one or two settings ranging from child guidance clinics to departments of psychiatry.

Interest in family therapy is growing as the result of work in several major centres in this country and abroad. The bewildering state of the art stems from different notions of what the outcome of therapy should be and how families are able to change. It is clear that the therapist's own impact on the family he treats, from his manner, to his personality, to his system of beliefs, plays a crucial role in determining outcome. The field is populated by charismatic figures whose personal influence cannot be overlooked in assessing the effectiveness of treatment.

Today family therapists can be roughly grouped into two. One considers the present cannot be adequately understood or treated unless the past is understood, while the other feels understanding of the past is not necessary, and that what matters is changing the ways families operate in the present.

Exponents of the former include Robin Skynner in the United Kingdom and Murray Bowen in the United States; and of the latter, Salvador Minuchin in America and Mara Selvini-Palazzoli in Italy are two leading figures.

For an up-to-date account of the similarities and differences between the schools of thought as well as a history of the family therapy movement, Vincent Foley's *An introduction to family therapy* (1986), is both concise and readable. For the general practitioner it is important to be aware not so much of the writings of any one individual as of the range of alternative therapies that exist. Hopefully, as comparative studies emerge, it will be possible to be selective as to what types of therapy suit particular family disturbances—but that is a long way off.

10 Working with individuals

The individual is both the starting point and the end point when it comes to dealing with family issues. It is the individual who often holds the 'illness' for the rest of the family; it is he who needs to be relieved of his burden of symptoms.

Because an individual needs to be seen in a wider family context and because the family can assist in the process of recovery, seeing family members together is important. Engaging others in the task of understanding someone's difficulties and working towards a common goal is desirable, but sadly not always possible. Many people live apart from their families. Others show a distinct reluctance to involve those close to them in what they consider to be problems unique to themselves. There are topics which individuals are ashamed or frightened about exposing in the company of others, yet are prepared to share with their doctor in the privacy of the surgery. Finally, there are problems which, though rooted in the family, make their impact felt particularly in one individual whose special needs cannot be ignored. Such vulnerable people do merit our special attention, especially at times of crisis. Even if we cannot have access to the rest of the family, there is every reason to offer help to these distressed individuals.

As with family work, helping individuals with family problems demands clear objectives. Rather than aiming to bring about radical change, the doctor may decide to allow the individual patient space to ventilate difficulties, to encourage clearer thinking about the underlying issues involved, and to give time to come to terms with problems that can only be successfully eliminated by involving other members of the family. It is still necessary to gain an accurate understanding of the particular problem and needs of that individual within the family. It is also imperative to come to an accurate assessment of that person's mental state. Perceptions can be coloured by depression and anxiety and occasionally by psychotic illness. We should not abandon our reliance on the traditional medical model of psychological dysfunction, since it may lead to interventions that are beneficial, for example, an admission to a short stay psychiatric unit when faced by life-threatening depression; or psychotropic medication for those severely disabled by lack of sleep, impaired appetite, and circular depressive thinking. Equally we should be ready to accept that many disturbances of psychological functioning are not the result of underlying defects, but of a temporary inability to cope with an overwhelming family-based stress.

Table 10.1 *Crisis theory and the medical model*

	Crisis theory	Medical model
PROBLEM:	'Crises' (emotional reactions to situations in which normal adaptive coping mechanisms are overtaxed by overwhelming life stresses).	'Mental illness' (underlying biological or characterological defects which result in manifest symptomatology).
AIM:	Help individual experiencing crisis cope more effectively with current stresses and develop greater capacity for future coping.	Diagnose and treat underlying illness.
METHODS:	Short-term crisis intervention.	Traditional psychiatric diagnostic and treatment procedures (diagnostic interviews and/or testing, dynamic psychotherapy, hospitalization, psychotropic medication, electro-convulsive therapy, etc.)
GOAL	Restore individual to pre-crisis level of functioning and render him less vulnerable to future stresses.	Cure illness and restore individual to a state of 'mental health'.

Source: Ewing, C. P. (1978). *Crisis intervention as psychotherapy*. Oxford University Press.

CRISIS AND THE INDIVIDUAL

Many situations at home are able to threaten the well-being of an individual: illness, the loss of a loved one, either in the immediate or extended family; the addition of a new family member; any major threat to a person's health within the family; or simply a move from one stage in the life-cycle to the next. Where the normal methods of tackling problems and stresses fail, and where the emotional integrity of the individual is threatened, a state of 'crisis' is said to exist. 'Crisis' is not itself a pathological state but rather a struggle to adapt in the face of difficulties which at the time seem insuperable. Crisis theory stands in direct contrast to conventional illness theory as Table 10.1 shows, since psychological distress is not seen to be symptomatic of illness but a normal consequence of stress.

Helping an individual can proceed along two distinct lines: one by diagnosis and treatment of 'emotional illness'; the other by helping that individual develop more effective ways of coping with the impact of family stress—so called 'crisis intervention'. The main difference between them lies in the extent to which the doctor assumes responsibility.

In the case of someone whose mental state gives rise to concern, it becomes necessary to look after the patient in ways which may be therapeutically helpful, but which foster dependency. Where the capacity to cope is threatened, but where, too, the patient is in touch with his feelings and able to take on the responsibility for his own actions, dependency should be actively discouraged and instead some form of crisis intervention offered.

Crisis theory was originally developed by Lindemann* (1944) and the concept was later elaborated by Caplan (1964) to provide a basis for helpful intervention at times of stress. It encompasses several techiques which have largely developed independently of each other to meet the needs of particular situations. They do, however, share a number of features:

1 Intervention needs to be swift in the face of crisis.
2 Treatment should be short-term and usually have time limits set.
3 The focus should be on current problems rather than past events, particularly on what has caused that person to seek help then.
4 The aim should be to help the patient with current difficulties, but also to develop a way of coping with future problems and crisis.

The doctor has to help his patient to face the realities of his current situation. It is permissable to give advice and make suggestions to encourage better understanding and to offer a means of dealing with feelings.

It is also acceptable for the doctor to show feelings and reactions himself; not merely to reflect in an impassive way on what the patient says or does. This openness on the part of the doctor encourages the patient to re-examine his own ideas, challenge his understanding of them, and make him assume some responsibility for changing them. Many patients are helped by this approach; the few that are not can often be pointed in the direction of longer-term psychotherapy, crisis intervention being a stepping-stone in this direction.

Every crisis increases the risk of psychiatric illness, but it also offers the potential for personal growth, a point worth stressing. The notion of making a fresh start can help in recovery since it provides a sense of hope.

* Lindemann, E. (1944) Symptomatology and management of acute grief. *American Journal of Psychiatry*, **101**, 141–8. Caplan, G. 1964. *Principles of preventive psychiatry*. Basic Books, New York.

CRISIS AND THE REST OF THE FAMILY

Crisis usually leads individuals to turn to others for help—not just to the doctor but family and friends. In this sense, a crisis is rarely experienced by an individual alone, and this can lead the doctor to see more individuals than just the patient. It is worth bearing in mind that families, like individuals, are more susceptible to outside intervention at times of crisis than at times of stable functioning. A family in crisis is less resistant to change than it is under normal circumstances. The general practitioner in caring for a family in crisis has a unique opportunity for dealing constructively with the family as a unit as well as with those individuals who present separately. It provides him, too, with the chance of anticipating and preventing problems amongst members of a family as well as helping him recognize and manage problems as they occur: consider the opportunities provided by visiting sick or elderly persons at home for recognizing and so influencing the mental health of those who care for them.

PROBLEMS IN WORKING WITH INDIVIDUALS

It has to be said that working with individuals on family problems can be fraught with difficulties. We are conditioned by our medical training to focus predominantly on the sick patient, and our first loyalty is to the individual who seeks help rather than the family who does not. He expects a diagnosis, and either advice or treatment for his problem. When this is not forthcoming, he feels disappointed, even annoyed. The time factor makes it difficult to tackle many important issues—ten minutes is barely enough time to uncover deeper family problems, let alone to deal with them. Even when family problems are undisguised, there may be forces within the family which encourage the patient's continued 'sick' role. It is easier to have a sick member of the family to nurse than to confront painful issues which apply to the whole family. Failure of the individual to respond to help from the doctor may indicate sabotage at home; an undoing of the progress that has been made in the surgery.

Progress with individuals can be impaired, too, by family secrets. They pose a threat to family stability and often embody the very problem that lies at the heart of an individual's distress. By remaining undisclosed they can act as an impediment to progress. If shared between just one family member and the doctor, they can likewise be damaging. It is ironic that the very strength of the therapeutic relationship between doctor and patient—the sense of trust and privacy that exists in the consulting room—can, at times, be a weakness in helping people get better. To betray individual confidences would be unethical and damaging to the relationship between doctor and patient: yet, to allow certain secrets to exist and, by implication, be condoned by the doctor, may actually perpetuate problems. A husband may confide in his GP about an extra-marital affair, yet be prepared for his wife to receive treatment for her 'nerves' from the same doctor. A wife may feel able to discuss a past abortion with her doctor, but not with her husband who is mystified by her attitude towards sex and blames himself for the difficulties in their relationship.

Whilst every encouragement should be given to patients to reveal confidences to others close to them, if they remain resolute in not wanting these discussed outside the doctor–patient relationship, we cannot overrule their wishes.

PROMOTING INDIVIDUAL SELF-ESTEEM

It is an encouraging fact for the doctor that most people who live through stressful experiences do *not* become depressed, displaying instead varying degrees of resistance. This issue of individual susceptibility has lead to a search for 'protective' as well as 'vulnerability' factors in depression. Attention has been focused on the family as one of the major influences on a person's health. The observation that family ties both bring support and create stress suggests for instance that in the care of depressive illness it is not the mere presence of family that counts, but the quality of the relationships between family members. Having a high degree of self-esteem makes successful coping more likely, while a low self-esteem imparts a sense of hopelessness which can predispose to mental breakdown.

The American therapist, Virginia Satir, who linked low esteem to problems in family communication, had as her therapeutic approach, an emphasis on teaching clearer communication patterns. This, she argued, actually enhances individual self-esteem.

Apart from recognizing those with particularly low self-esteem and assessing their levels of depression and suicidal risk, our aim should be to restore some of the confidence and feelings of self-worth which have taken a battering. It is obvious that a belief in one's own abilities makes it possible to deal with problems as they arise and from whatever source. By focusing on areas of individual achievement rather than areas of failure, and by adopting a positive approach to people's difficulties, we are in a position to restore a sense of balance to their lives so that they feel able once more to tackle difficulties at home.

11 Working with couples

While marriage can satisfy people's deepest yearnings, it can also act as a constraint and a cause of much unhappiness. It occupies a central position in the system of family relationships and because of this the effects of marital harmony and marital disturbance are felt throughout the family.

The capacity of married couples to deal with conflict in their relationships should not be underestimated. Many would argue that a marriage free of conflict does not exist and that since conflict and personal growth are closely linked, the institution of marriage offers potential for psychological development which is difficult to match elsewhere. The extent to which couples achieve growth very much depends on what they bring to the relationship and, in particular, what they have learned from their primary attachments. Having a secure and loving relationship with one's parents helps, while failure of these early relationships renders people vulnerable to marital breakdown.

THE PRESENTATION OF MARITAL DISTRESS

It is the experience of many general practitioners that more people present their marital problems directly than was the case twenty years ago, although it is still the case that most marital problems are presented in disguise. Many are presented at a late stage in their evolution when couples are on the verge of separating or going through a trial separation.

The indirect presentation of marital stress requires the doctor to be vigilant for early signs in his patients. The forty-year-old executive who delays his journey home at the pub, the housewife with a history of pelvic problems, and even the child with abdominal pain can all act as pointers to marital disturbance. As the GP is in the front line when it comes to dealing with illness and behaviour disturbance, it is worthwhile taking an adequate history of relationships at home.

Whether or not couples venture with their problems to the doctor's surgery depends largely on their past experience—how they were received over other difficulties, how much interest was shown in social and psychological aspects of health, and how unhurried the doctor appeared to be. It is with considerable misgivings that many couples relate their marital problems to the doctor. For many it feels like an admission of failure—hence the importance of a sympathetic ear and an enabling manner.

THE DOCTOR'S RESPONSE

The way couples present their marital difficulties or symbolize their 'disease', determines the way the doctor responds. If, for example, a couple are anxious to examine the physical aspects of their distress—but less keen to discuss their psychological and social implications—the doctor has to confine his attentions to the symptomatic relief of distress. He may be disinclined to examine the deeper causes of unhappiness and may collude with the couple by not discussing the marriage at all. With another couple he may find himself flooded with marital unhappiness, both partners seeking his support against each other, or desperately turning to him for solutions now that other repertoires have been exhausted. To be a receptacle for so much feeling can at first be confusing and burdensome. What happens subsequently depends on his relationship with the couple and on their feelings towards him. The doctor–patient relationship can be a powerful agent for change. Realizing this allows the doctor to place more emphasis on this aspect than on the prescription pad.

DIFFICULTIES TO BE FACED

Time is a major constraint and there is little alternative to devoting extra time to couples who are in great distress. Confidentiality is another problem. As with all family matters, the general practitioner can unwittingly be the repository of confidential information from one party and must have the patient's permission before divulging information which could be upsetting to the spouse. Finally, like any other human being, the doctor is instinctively drawn to some individuals and not to others, and this, too, can affect his objectivity and performance.

ADVICE FOR THE BEGINNER

What general advice can be given to doctors who wish to embark on marital work?

First, marital problems are by definition *shared* problems. While individuals may have particular need which demand his attention, the main focus should be on the relationship. This means paying attention to what goes on verbally and non-verbally between the partners in the consulting room. It means encouraging a dialogue between the couple and seeing that comments are not exclusively channelled through the doctor, but finding their appropriate target and eliciting a response which can be examined.

Second, it is important for the doctor to remain impartial. Some individuals when faced with mixed feelings succeed in splitting off the bad ones and project them instead into their partner. On occasions the division is total, so that the partner becomes the culprit while the other holds all the positive feelings. In reality it is rare for the 'fault' to lie with one person. To apportion blame should be recognized as unhelpful. Attempts to win over the doctor's sympathies at the expense of the other should be resisted. When one partner is absent from the consultation it becomes even more important not to jump to conclusions about the absent spouse. All couples have a basic need to be understood, and this means listening to both sides of an argument and actively searching for common understanding.

As well as remaining impartial, the doctor needs to create a safe environment in which couples feel free to air their innermost feelings without these getting out of hand. Trust in the doctor is crucial and the rules of confidentiality may need to be spelt out so that there is no suspicion that what is spoken will go further. Any records made may need to be kept separately from the main notes, even given to the couple themselves for them to digest before the next session.

THE CORRECT APPROACH

What route should the doctor take once he has been successful in engaging both partners in the task of working on their relationship?

'Marital therapy' is a term which encompasses a variety of different theoretical models. Two main approaches, behavioural and psychodynamic, need to be considered.

Behavioural methods do not concern themselves with promoting insight. They ignore feelings and the roots of conflict. They are not concerned with the meaning behind communication. Their primary focus is on behaviour in the 'here and now'. Despite their simplistic view of human behaviour, their effectiveness can be demonstrated in many cases.

Communication training is one behavioural approach to marital problems which lends itself to general practice. Many couples have lost the art of discussion. Either their conversation becomes too heated or they degenerate into a list of complaints about the other. Opening up paths of communication, pointing out the destructive aspects of what is said, and substituting less destructive words can be useful.

A **psychodynamic** approach to working with couples lends itself particularly well to general practice, making use as it does of the relationship between doctor and patient. One such approach has arisen out of work done at the Institute of Marital Studies, in London.

Four concepts underline much of the thinking which has emerged from this organization:

1 Many problems in marriage stem from people's infantile and childhood experience.
2 Unconscious as well as conscious factors play an important part in bringing couples together, in sustaining their relationship, but also in creating problems within the marriage.
3 Conflict in marriage is an attempt to redress, in the present, conflicts which have arisen in the primary relationship.
4 Not all conflict is bad, since without conflict there is no personal growth.

Working with couples in a psychodynamic way aims at bringing about a change in the way partners see themselves and each other. The approach stresses the importance of emotional growth and self-awareness.

In adopting a psychodynamic approach right at the beginning, it is helpful to ask oneself three questions:

1 *Why did this couple present their difficulties at this time?*

The gestation of difficulties can be a long one, but there is usually some event or change in circumstances which prompts people to seek help. It is important to know what this is and what threat is poses to the marriage.

2 *Why did this couple choose each other in the first place?*

Although consciously partners select each other for their 'good' and positive aspects, at an unconscious level it is because they are at the same stage as each other emotionally. Depending on their relationship, and whether or not further emotional growth has occurred, they either stay together or develop separately. What unconsciously they shared at the time they met and what happened to each separately can be helpful in understanding the problem.

3 *What is being expressed by one for the other?*

Mention has been made earlier of the situation in which a couple divide things up between them. One may carry all the negative aspects of the relationship while the other hangs on to the more positive, 'healthier' aspects. One is essentially using the other as a container for certain unwanted attributes. Seeing what aspects of a person cannot be tolerated, or tolerated only when projected into another person, is a necessary prerequisite for growth. If all the difficulties are projected into the other partner, the relationship may be kept 'safe', but it is sterile from the point of view of growth. In psychodynamic marital therapy an attempt is made to redress this balance, so that the disowned parts of the person are re-incorporated. This means helping people to tolerate aspects of themselves of which they are largely unaware. In many cases the partners remain split as individuals and divided in their relationship with each other, resisting attempts to change this balance.

The choice as to whether to work behaviourally or psychodynamically with a couple depends both on the doctor (what he personally feels most comfortable with), and on the couple (whether they employ defences against, or actively struggle with developmental issues).

SOME EXAMPLES

Finally, let us consider three couples and possible approaches to their problems.

Couple A are full of recrimination and self-blame. They show concern about the way their unhappiness is making their child's life miserable. Each is preoccupied with his or her role in the disturbance. Knowing something is wrong is a long way from understnading it and, unless the parents examine their own interaction, it may be difficult to bring about the improvement that both of them desire for their child. Here a psychodynamic approach has most chance of succeeding, given the willingness of the couple to confront their painful feelings.

Couple B spend much of their time voicing their differences and resentments about each other's behaviour. Each blames the other for what has happened. Accusations fly but, just as they find it impossible to live together peacefully, they find it difficult to separate. Here the doctor is put in a position of stalemate. Progress seems impossible and the situation is uncomfortable for all parties. The frustration engendered in the doctor makes it necessary to share the dilemma with another: conjoint marital therapy may be the answer. With two trained therapists symbolizing parental figures, the couple may be able to contain their feelings and avoid the processes that make them locate the problem in another person.

An alternative behavioural approach relies on opening up new paths of communication which are less destructive. Crowe lists some examples of common destructive communications and provides us with suggestions for feed-back (see Tables 11.1 and 11.2).

Table 11.1 *Examples of destructive communications*

When the couple talk to each other, it is almost inevitable that one or other will make a comment that intrudes on the sensitivities of the other. Such comment may be of many types, for instance:

1 Destructive criticism: 'You're a real coward' or 'You really like to hurt me'.
2 Generalizations which often amount to character assassination: 'You're always showing me up' or 'That's typical of your immature attitude'.
3 Mind-reading, or imputing destructive motives: 'You say you're trying to help, but I know it's just to humiliate me'.
4 Bringing up the past (often combined with generalization): 'I've suffered this for

16 years and you're still doing it' or 'I'll never forgive you for what you said to my mother'.

5 Putting oneself in the right and partner in the wrong: 'I've never started a row in the whole of our marriage' or 'All I want is peace, and look at the way you treat me'.
6 Using logical argument as a weapon: 'All you have to do is to stop overreacting' or 'Be reasonable, can't you see that this is the only way?'
7 Raising the voice: a ploy that can easily be overlooked.
8 Using the sting in the tail: 'You've stopped criticizing me, but now you've gone silent and that's worse', or 'You are trying now, but why couldn't you have done so five years ago?'.

Source: Crowe, M. (1982). The treatment of marital and sexual problems. (Chapter 13) Vol. 1. *Family therapy*. (ed. A. Bentovim, G. Gorell-Barnes, and A. Cooklin).

Table 11.2 *Examples of beneficial feedback*

Here are some examples of the kind of feedback which can be given, both in terms of teaching of principles and the actual suggestions for phraseology:

1 Use expression of one's own feelings instead of imputing feelings to the other partner: 'You really like to hurt me', becomes 'I felt very sad to hear you say that'. This example has three improvements: the speaker's feelings are expressed, no imputation of intent is made on the speaker's side, and the sentence applies to a specific and not a general situation.
2 Use specific examples: 'You're always showing me up', becomes 'I felt really embarrassed when you spoke about me to George last Friday.'
3 Avoid mind-reading: partners are requested to ask about motives rather than to insist that they know better than the other partner what he is thinking.
4 Stick to the point: this can mean not bringing up the past, and not generalizing about similar instances.
5 Replace logical arguments with expressions of ones own emotions: 'All you have to do is to stop over-reacting', becomes 'It makes me feel hurt when you raise your voice'.
6 Therapists can point to non-verbal aspects such as body posture or eye contact as something that raises sensitivites.
7 Partners should be encouraged to take responsibility for their own actions: 'You always make me angry and then I get violent', becomes 'I can't seem to control my temper'.
8 Often monologues kill communication, and these can be politely discouraged, with a clear message that short, sharp comments are more desirable (note the phrasing of this sentence, with emphasis on the desirability of one rather than the undesirability of the other).

Source: Crowe, M. (1982). The treatment of marital and sexul problems. (Chapter 13). Vol. 1. *Family therapy*. (ed. A. Bentovim, G. Gorell-Barnes, and A. Cooklin).

Couple C present with sexual difficulties, such as an absence of desire for any sexual relationship. Although problems of intimacy extend beyond the sexual into the emotional sphere, the request for help is specifically for sex therapy.

Here is not the place to review the literature on sex therapy. The general practitioner has access to specialized literature such as Masters and Johnson and Keith Hawton's book. It is important, though, to realize that sexual problems both draw attention to sensitive areas in the marriage and serve to deflect attention away from other interpersonal aspects of the relationship. Couples like this one who isolate the sexual complaint as a problem may, as Clulow (1984) suggests, be indicating to their doctor that their anxiety lies elsewhere, but, at the same time, have a vested interest in keeping the problem alive as a protective device.

THE CHALLENGE OF MARITAL WORK LIES IN FINDING AN APPROACH WHICH DOCTOR AND COUPLE FEEL COMFORTABLE IN HANDLING.

Several agencies exist which offer help and support. Issues which arise out of sharing cases or referring couples are discussed in Chapter 13, and information about helping agencies is given at the end of the book.

12 Working with parents

A TIME OF MUTUAL NEED AND ADJUSTMENT

Not enough attention is paid to the needs of parents.

It is axiomatic that the child comes first, but the needs of the children are inextricably bound up with the needs of their parents. A problem for one is, by definition, a problem for the other. We assume too readily that parents are competent to manage, and motivated to act upon, problems which stem from their relationship with their children and with each other. True, parents are the real providers of primary care for their children, but it is a particularly demanding job, and one in which even the most capable and resourceful parent requires help and support.

A measure of this need comes from the degree of contact that exists between mothers of infants and members of the primary care team.

In a study in Kentish Town (Williams *et al.* 1981), the number of

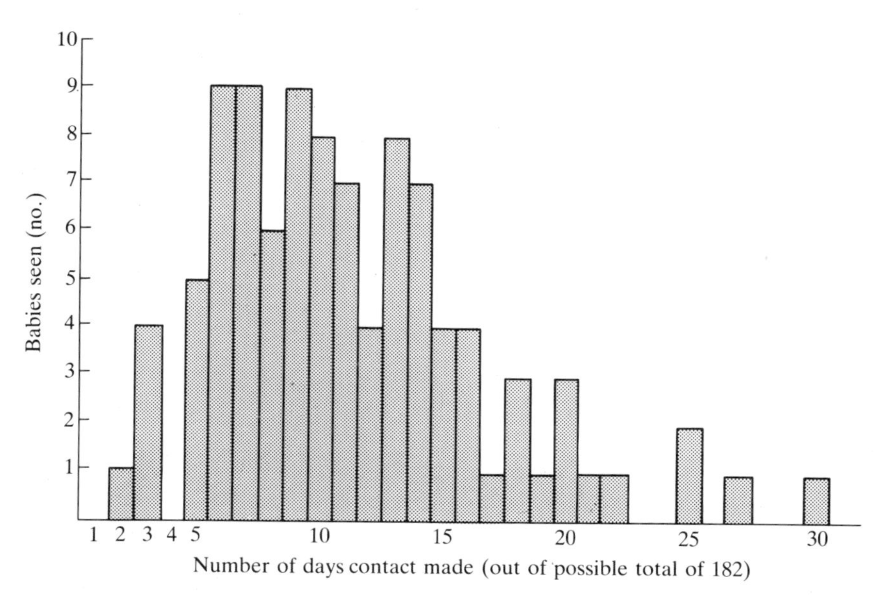

Fig. 12.1 Number of babies seen in relation to number of contacts. Mean = 11.2, Median = 10.3. Williams, P. R., Argent, E. H. M. and Chalmers, C. (1981). A study of an urban health centre: factors influencing contact with mothers and their babies. *Child care, and health and development*, **7**, 252–56.

Table 12.1 *Number and percentage of contacts over the study period.*

Contacts	Number	per cent
Baby clinic	523	46
Doctor at home	54	5
Doctor in surgery	182	16
Health visitor at home	256	22
Health visitor in health centre	32	3
Health visitor over telephone	73	6
Practice nurse in health centre	26	2
	1146	100

Source: Williams, P. R., Argent, E. H. M., and Chalmers, C. (1981). A study of an urban health centre: factors influencing contact with mothers and their babies.
Child care, health and development, **7**, 255–6.

Table 12.2 *Average number of home visits by the health visitor taking acount of maternal mood and whether the child was first-born.*

Mood	First-born	Not first-born
Depressed	2.55	1.70
Not depressed	1.64	1.09

Source: Williams, P. R., Argent, E. H. M., and Chalmers, C. (1981). A study of an urban health centre: factors influencing contact with mothers and their babies.
Child care, health and development, **7** 255–6

contacts made by mothers in connection with their infants during a six-month period was examined. Amongst the hundred families studied, over a thousand contacts occurred, half of which were in the Child Health Clinic. Of contacts outside the clinic, 58 per cent were with health visitors, the majority being in the parents' home. A high level of contact was maintained in a wide variety of situations, and there was no indication that a higher than average attendance at the Child Health Clinic, with its emphasis on prevention, had a sparing effect on the use of the doctors' and health visitors' time in other situations. The degree of help sought appeared disproportionate to the physical problems encountered. Why then this high level of contact? Different reasons accounted for this in different settings. For example, being a first-time mother with symptoms of depression added significantly to the likelihood of a visit from a health visitor, while living within a mile radius of the health centre and having more than one child increased the likelihood of a visit from the doctor. It seemed likely that the most important determinant of contact overall was a need on the part of the mother for reassurance and support, something not always forthcoming at home. Moreover, on several occasions it seemed as though the child was acting as an admission ticket for the mother who herself needed help. Analysing tape consultations between mothers and doctors was revealing. Mothers frequently commented on their own well-being and used the opportunity to speak on matters affecting, not just themselves, but other members of the family. It seemed as though mothers wanted to broaden the agenda for discussion beyond the immediate concern with the child. Family problems were frequently uppermost in their minds.

Findings such as these need substantiating since they may be peculiar to inner city practices. Nevertheless, they strike a familiar chord in many general practitioners' minds.

Parenthood is a time when adjustments have to be made. Family problems often surface with the arrival of children. While a great many families negotiate the transition without outward signs of difficulty, many seek help over child management problems, particularly sleep and feeding disorders; they appear to have a lower threshold for consulting over apparently 'minor' ailments. Having children is a fearsome responsibility. Young children are unable to articulate what is wrong with them. Help is needed in making an accurate diagnosis. Some children actually signal through their symptoms problems of relationship, not just between mother and child, but between mother and father. For this reason it is important that the GP and health visitor, as first recipients of children's complaints, are attuned not only to the distress of the child, but to that of the parents too.

In extreme cases parents, rather than face the trauma of separation from each other, route all their conflict through the child. They may stubbornly refuse to acknowledge any trouble in the marriage and use the child as a pawn in their conflict avoidance. Ironically, the more they focus on the child's symptoms, the worse it becomes, until it forms part of that family's way of life. When marital stress has been routed through the child, and, if for some reason the child is removed from the family, it is common for one or other parent to fall ill. This 'retreat into illness' may even be encouraged by the other partner, since again it provides a convenient excuse for not facing up to the painful issues in the marriage. Such parents can be very difficult to help and the child who is caught up in the conflict can be left with permanent emotional scars.

THE FAMILY CONFERENCE

In less intractable cases, and where the parents use their child as a ticket for gaining entry to the doctor to talk about themselves, a family conference is invaluable. By appealing to both parents in their joint role as guardians of their child, and by stressing the demanding nature of parenting, they can be made to feel less inhibited about talking over their difficulties and encouraged to talk about inward concerns. Not uncommonly, conflict between husband and wife over parenting issues emerges. Whether or not this speaks of conflict elsewhere (for example, in the marriage), it is important when dealing with it to focus on areas of agreement in parenting rather than on areas of disagreement. Some couples will not want to do this, offering instead excuses for not collaborating, or retreating into generalities rather than face specific managment issues. It is important to be firm and to say that without agreement progress is impossible. Another principle is to build on parental strengths rather than dwell on weaknesses. In dealing with differences of opinion, it does no harm to stress the danger of inconsistency towards the child, and the benefits of adopting a shared strategy in managing the children's problems. Even when parents themselves are in bitter disagreement in their own relationship, they should be encouraged to come together in the interests of the child, so that they share the responsibility for parenting and avoid using the child as a weapon in their conflict.

LOW MORALE AMONGST PARENTS

In the course of a family session, where morale seems to be low, special emphasis needs to be placed on the needs of the mother. Frequently maternal depression itself is the most important underlying factor and merits separate attention. Feelings of hopelessness and a sense of isolation are common amongst mothers of young children. Whether this is consequent upon a poor marriage, bad housing, or stress from some other quarter, there is an increased risk not just to the mother but to the children as well. Specialized help may be required, even admission to a mother and baby unit if this situation demands it. GPs themselves should not feel backward about offering time as well as anti-depressants to those mothers who continue to manifest a depressed mood, sleep and appetite disturbance, and an absent libido.

Confidence in parenting is lost not only as a consequence of depression but in situations where the parent has no-one to turn to for help and support. This particularly applies to single parents and to families where one partner is unavailable either through illness, a demanding job, or because of problems in communication. Ethnic minority groups feel additionally cut off because of cultural and language difficulties which result in further social isolation.

THE NEED TO INVOLVE FATHERS

One task which commonly faces us in practice is the need to involve fathers more in the work of parenting and in supporting their wives in their jobs as chief care-giver. Fathers withdraw for a number of reasons. Some feel that looking after children is exclusively women's work. Others simply do not have the time to devote to their families that they would like. A proportion feel excluded as though displaced from their wife's affection by the arrival of successive children. They all need to be encouraged back into the family if they are not to drift into substitute relations outside the home or become permanently estranged from their wives and children. The rationale for involving fathers has much to do with preventing maternal depression and its consequences for the children as with sharing the work of parenting. Many families with young children break up unnecessarily because of a poor understanding of each other's needs. At the extreme end of this range of isolation are the parents who live alone, lack kinship support, and who have a history of marital violence or come from a violent and chaotic home background. These are the parents who are likely to harm their children.

THE VALUE OF CHILD HEALTH CLINICS

Fortunately most parents have some form of help, as well as access to finance to support their children; they understand welfare services sufficiently to call upon assistance when in distress. Warning signals can, however, go unheeded and here we need to realize the special function of child health clinics. They provide not just a convenient meeting place for mothers and health workers, but a route by which those in particular need of support can seek help both for their children and for themselves. The popularity of such clinics speaks for itself. Much more can be done within the context of the weekly clinic to help parents in difficulty.

1. Child-psychotherapists

Knowing that parents and children can be helped by brief psychotherapeutic contact with professionals, attaching a child psychotherapist to a clinic can have great advantage.

One such therapist (Dawes 1985), described how she would place herself centrally near the weighing area, a focal point in any baby clinic, so that she was visible and available to parents, doctors, and health visitors alike. From this position, she was able to respond to problems as and when they arose. Since she was also able to see a wide range of mothers and babies interact, her impressions became important on occasions when intervention was considered necessary by the doctor or health visitor. She would participate in discussions at the end of the clinic when attendance was reviewed and her observations were often helpful in recognizing and in understanding the problems that were presented there. More often than not, she provided in discussion a fresh view of a problem rather than an opinion arising out of a separate referral (thugh she did see parents and children, too). In the truest sense of the word she acted as a 'consultant' to the clinic.

2. Self-help groups

Apart from involving psychiatrically trained staff from outside, child health clinics can be made more effective in meeting the needs of parents through self-help groups. These can be run by health visitors alone or in collaboration with doctors. The improved communication and understanding which is generated is not only evident in the group sessions but at other times and in other situations where contact is sought with professionals.

ADOLESCENCE AND ITS PROBLEMS

The problems of young parents are matched by the problems of parents with adolescent children. Whether one regards adolescence as a normal developmental phase or a time of crisis, the family as well as the adolescent have to contemplate an important set of personal changes. The likelihood of things going wrong is increased if the family itself is in disarray. This frequently happens when husband and wife are at loggerheads, or when the adolescent is part of a reconstituted family. As is often the case with the younger child, for the adolescent to 'get better' when he is the symptom-bearer for the entire family, constitutes a threat to other family members, who therefore strive to keep him in the 'sick role'.

The adolescent is himself going through a painful transition. Like a younger child, he may be unable to articulate his needs except as behavioural outbursts or physical symptoms. Abuse of alcohol and drugs can be added to this repertoire. The doctor moreover may be cast into the role of yet another authority figure (like parents), so that he is the recipient of a lot of anger and mistrust, or barred altogether from helping. Where the adolescent is part of a system of conflict avoidance in the parents, he can again be difficult to help. Shifting the focus onto the marriage often meets with resistance, and the marital problems may only be acknowledged when the unacceptable behaviour on the part of the adolescent has stopped.

Despite these drawbacks it is often worthwhile to bring the family together to discuss their difficulties. To work just with the adolescent on his own can be especially difficult, since progress is often sabotaged at home by others in the family who offer resistance to change. Whether one uses a psychodynamic approach in dealing with the family or a behavioural approach (so that changing the behaviour of family members becomes a priority rather than promoting insight), is a matter for each individual doctor. Success can be difficult to measure, but the general practitioner who has the opportunity to watch the behaviour of his patients over years is in the best possible position for assessing interventions. Experience alone will help him to decide which families respond best to help, and which method suits both him and his patients.

13 Working with other professionals

Knowing how much anxiety and uncertainty is generated in the doctor by working with complex family problems, it is hardly surprising he should feel the need to share this with others. While doctors are accustomed to having feelings like depression and anxiety projected on to them by their patients, they also have need of defences against the full impact of such emotions. Assuming a detached position is one form of professional defence; others are sharing the case with another professional or referring the family on for further help, thus extending the responsibility for management.

Involving a third party invariably disturbs the relationship between doctor and patient and not infrequently meets with strong resistance. In the case of couples who find it difficult to handle their ambivalent feelings for one another, the resistance to referral often comes from one party who carries all the fear and anxiety for the other. For the doctor, too, there can be obstacles. He may be reluctant to relinquish care to another professional, feeling that it may disturb what has hitherto been an exclusive relationship. The case will often be taken out of his hands. He may, in addition, experience a sense of personal failure.

On other occasions the doctor recognizes a strong urge in him to be rid of the case altogether and, when the opportunity for referrral arises, he is only too anxious to comply without thinking of what will be achieved. In every case it is worthwhile asking what is the *aim* of referral and what are the *expectations of* the person or agency to which the case is referred.

In the case of family problems, referral usually means involving some branch of the psychiatric service, though with closer links being found between GPs and community-based workers such as marriage guidance counsellors, social workers, clinical psychologists, psychotherapists, and health visitors, help will sometimes be nearer at hand than the local hospital. Much is gained by having help immediately accessible preferably sited in the same building. Seeking assistance some distance away, when contact is limited to the exchange of letters, often without the benefit of a name, can be fraught with difficulty. Where professionals do not know each other and have a poor idea of the way each functions, the scene is set for misunderstanding. Moreover, differences of opinion between agencies are often exploited consciously or unconsciously by families or couples. The conflict within the family becomes an aspect of the behaviour of the professionals and what is acted out by the helpers is often the very conflict that brought the family to the doctor in the first place. When more than one agency is involved with a family, it becomes imperative that communication between them is easy and clear so as to prevent splits and misunderstandings from interfering with treatment.

USEFUL LIAISONS

The contribution that can be made to general practice by the attachment of a psychotherapist has been examined by Brook and Temperley (1976). The purpose of the liaison was to increase the resources of the practice and reduce the amount of splitting which occurs in patient care. In both respects the attachment was a success. It was noticed, however, that the working relationship between professionals often became infiltrated by the emotional conflicts of the patients they were helping. This would surface in the course of discussion between workers in their weekly mutli-disciplinary workshop. Rivalry between professionals also occurred and appeared to be at the root of some intractable difficulties encountered. Once this was acknowledged, a new perspective on the work could result. Difficulty in sharing the care-giving role was sometimes expressed as an over-eagerness on the part of the doctor to withdraw from the case, allowing the psychotherapist the opportunity to demonstrate 'superior skill'. Despite these problems, the experiment seemed to increase doctors' ability in identifying, tolerating, and alleviating psychological distress in their work, and it was also of benefit to the therapist who learned something from the attachment to a group practice.

Graham and Sher (1976) reported on a similar practice attachment in which a social worker with psychotherapeutic skills paid regular visits to an urban group practice in London. The outcome of their collaborative work was to free the general practitioner from some of the frustrations involved in coping with families and individual patients. It was noted how GPs often had to resist enormous pressures from the families of patients to label one of their kind as being psychiatrically ill and in need of intensive psychotherapy or immediate hospital care. Talking over the problem with someone with psychodynamic training helped the doctor to face the guilt and anxiety which arose out of the decision to refer or not, and helped, too, in allowing him to regain a clearer perspective on his relationship with the family. But the experiment was time-consuming and created unease amongst other members of staff who saw the creation of 'special patients' (ones who were likely to generate an extra burden of care), as being potentially dangerous. As in the previous study, feelings of rivalry and competition between professions were ever present and needed recognizing and talking through.

THE PROBLEM OF COLLABORATION

The benefits of collaboration in both examples quoted exceeded the disadvantages of working together. However, simply working closely with professionals does not ensure better communication. Close working relationships can, on occasion, result in worse communications. Discussion of a case can be relegated to a few moments in the corridor. Instead of spending time thoughtfully constructing a letter or referral and asking what is likely to be achieved by it, disjointed thoughts and anxieties on the doctor's part are aired in the hope that someone else can provide a solution to a difficult family problem. GPs have a highly developed sense of autonomy and share a culture of independence. Working closely with other professionals even when this is evidently in the interest of their patients does not come easily.

If problems of sharing casework with other doctors proves problematic, they are even more so when non-doctors are involved. In a recent study of the ways in which marital difficulties are managed, it was evident that GPs seldom communicated with non-doctors and likewise non-doctors rarely communicated with them. Woodhouse and Pengelly (personal communication) suggest that perhaps the most important reason for this was 'the outsider', be it a social worker, psychotherapist, or marriage guidance counsellor, posed a threat to the boundary around the exclusive relationship between doctor and patient. Effort was expended in preserving this boundary, so protecting the relationship from intrusion from other professionals. Much is invested in the doctor–patient relationship and the rules of confidentiality make it difficult to share casework. Yet, the advantage of sharing a common body of information and of achieving new perspectives on family behaviour are obvious.

The idea of co-therapy, bringing professionals from different disciplines together to form a therapeutic partnership in managing peoples' problems face to face, is an attractive one when dealing with family groups. Child guidance clinics have traditionally used psychiatrists to take responsibility for the treatment of the child, while social workers are responsible for helping parents. There are few examples of such partnerships within general practice. The reasons for this are not difficult to see. Both professionals need to trust one another and be accustomed to each other's style of practice. They need to function as a partnership with equal power and responsibility towards the family they are treating. They need to share the same aims in therapy, though there may be differences in perspective. Above all, they need to be alert to the power families have in splitting such a partnership, investing one with all the goodness and turning the other into a weak or persecuting force. It is not easy to achieve a comfortable sharing of power and those professionals who do engage in co-therapy need time to work on the difficulties inherent in their relationship. To date, there is no proof that co-therapy is any more effective than single person intervention in managing family problems. Faced with such a powerful system as a family, there is no easy answer to working effectively with others, except by investing time in understanding each other's role and being conscious of the ways families can create rifts between the workers who are trying to help them.

ACHIEVING CLEARER COMMUNICATION

One way of achieving this is by writing better letters. The referral letter remains the commonest form of communication that general practitioners have with professionals outside their practices.

In their dealings with others, general practitioners often fail to make their requirements clear. Instead it is left to the other party to guess whether it is advice on management which is required, or simply a take-over of the case. In turn, replies can be lengthy, full of jargon and be unhelpful to the general practitioner. To share the burden of a difficult family or even to be free of it for a while, the general practitioner should consider whether the following points are contained in any letter of referral.

1 Is what is required of the other professional explicit?
2 Does it entail important aspects of past history which the family may think unconnected to their present difficulties?
3 Does it answer the question as to why the referral is being made *now* rather than previously or in the future?
4 Does it mention contacts with other professionals who may already be involved with the case?
5 Is it clear whether or not the general practitioner wants continued involvement with the family over the problem, and if so, in what capacity, for example as prescribing doctor, source of support, or co-worker?

Such guidelines can set the stage for better collaborative work and so benefit all parties.

Knowing the person to whom one is referring can have obvious advantages. Even one face-to-face contact can free communication sufficiently for each to trust the other in later dealings with the family either by letter or over the telephone.

Finally, allocating time in the week for informal discussion of family problems with other members of the primary care team, for example following a Child Health Clinic, with the practice health visitor, or over a working lunch with one's partners or a visiting psychiatrist, acts both as a support to the doctor and offers the potential for a fresh perspective on the case. Working in this way with professionals should no longer be seen as a luxury but as a prerequisite for proper understanding and better management of family problems.

14 Prevention

VULNERABILITY AND RESILIENCE

In order to prevent or ameliorate family problems it is necessary to understand what makes some families vulnerable and others resilient in the face of disturbance. While illness and loss can cause considerable family upheaval, most families withstand a substantial amount of upset. Each family, however, has its threshold above which coping mechanisms fail and problems arise. The level at which this is set depends on a number of factors.

1. The exact nature of the disturbance

A sudden loss can have more impact on a family than one that is anticipated. Chronic illness carries its own legacy of difficulties—principally for the chief carer, but also for others in the family who may be required to take on new jobs and relinquish old ones. Even minor additional responsibility can lead to conflict in some families, so nice is the balance within the family system. Acute illness likewise can be upsetting particularly where other adversities operate, such as housing problems and financial difficulties.

2. The previous experience of families in terms of their exposure to illness and loss.

One loss can spark off memories of previous losses. In this way the timing of disturbance is influential in creating additional problems for families and individuals. Past experience also invests more recent events with particular meaning, which in turn governs the family's response.

3. The quality of relationships within families.

Clearly the more supportive family members are towards one another and the better the level of communication, the less likely it is that problems will become chronic or be dealt with inappropriately. The mere availability of friends and relatives is of less importance than the use people make of relationships at times of crisis.

4. The methods families employ to solve their problems

Many families show a willingness to work together constructively in solving their problems, accommodating to any changes that are needed. Others appear to run away from difficulties, employing instead denial or putting the blame for their misfortune on one person, the scapegoat for the rest.

5. The vulnerability of particular individuals within families

Some individuals, perhaps because of emotional illness or the timing of their birth, or just because they are young and immature, are rendered particularly vulnerable to family-based stress.

IMPLICATIONS FOR PRACTICE

What implications do these factors have for preventive work?

We are already concerned with minimizing the impact of illness as well as preventing it. Knowing the effect loss has on individuals and families, we are in a position to prepare them where death is anticipated and help those suddenly bereaved in facing their grief. Knowing the importance of previous losses we need to be alerted to these; this has implications for our record-keeping. The dates of important past events, death of close relatives, stillbirths, and other significant losses need to be entered on the summary page of the notes. Some practices use this information to contact relatives around the time of an important anniversary, but mostly it is used as an *aide-memoire* for consultation.

Where families are exposed to chronic illness and handicap, doctors can be active in supporting the carers both by offering practical assistance and emotional support. As the contribution close relatives make to the care of the elderly, the sick, and the handicapped becomes more recognized, there is need to sustain, promote, and at times substitute for it. Demographic trends point to fewer carers being available to meet the need in the future. The main need for sustained care comes from those over 75 years whose numbers are rising steeply; and yet the number of adult children available to do the caring is falling. One solution is to link formal and informal care. As it is, home helps, good neighbours, and nursing services assist in the caring for people at home. However, few dependent people who have informal carers receive these services, and then only in times of crisis. There is need for an integrated approach in which informal care is acknowledged more and the strain on relatives eased before problems reach crisis point.

While we cannot easily improve the quality of relationships within the family or prevent the breaking up of partnerships, we can, given the opportunity, open up lines of communication. The growing demand for help with marital difficulties speaks of a need for some third party to assist in helping to sort problems out rather than allow family break-up. It is not that couples have not tried to solve the problem themselves, but they may have exhausted their repertoire of solutions. Helping them to see more clearly what is hindering that process, and offering new ways of looking at the same problem, can be of great benefit in the future.

Finally, in the case of vulnerable members who are used as pawns in family disputes, the doctor can intervene as an advocate for those who are being manipulated by others and also help parents face the painful issues that cause them to route their anxiety and despair through their children.

SPECIAL OPPORTUNITIES FOR PREVENTIVE WORK

The distinction between 'prevention' on the one hand and 'diagnosis/treatment' on the other, is somewhat artificial. What is done about a family problem at the time it presents invariably influences what comes later.

At the price of repeating what has been said in previous chapters, it is worth emphasizing the opportunities which exist for preventive work in certain situations, namely during crisis, pregnancy, early parenthood, and at times of separation and divorce.

The preventive possibilities afforded by crisis

Crisis offers the real test of family functioning, since it is a time when people become vulnerable and hence more dependent upon one another. Although it raises the spectre of emotional illness and family breakdown, it also carries the possibility of change for the better. Previously it was novelists and playwrights who drew people's attention to the possibilities presented by crisis. Now psychiatrists stress how minimal intervention at a time of family crisis can produce lasting benefit in terms of the mental health of the whole family.

Helping to make a family master one set of difficulties helps them to achieve a higher level of competence when faced with difficulties in the future.

Though events which cause disturbance will vary, the approach people take to them may not change unless they become aware that a choice of response exists. Even when families continue to be exposed to the same set of problems, for example, where they are required to help a chronically disabled relative, helping them face up to and adapt more constructively to the situation can lessen problems for the future. This does not mean that the doctor is required to give detailed advice or offer lengthy therapeutic sessions. As Caplan (1961) points out, 'When the time is ripe, the right word in the right place gives better results than a lecture'. He also points out that some of the most powerful messages are often conveyed without words but through an understanding manner, a sympathetic nod or gesture, and by displaying patience.

Opportunities for prevention in pregnancy and early parenthood

In Chapter 12 the special part played by the child-health clinic and the antenatal clinic in detecting family difficulties before they became firmly rooted was discussed. Because disturbance takes time to become established, the general practitioner is in an excellent position to intervene early on and so prevent harmful long-term consequences for the family as a whole.

Pregnancy and the arrival of the first child are critical periods in family life when relationships are put under stress. The reverberations, like those from a loss, are felt throughout the family and have far reaching implications for future disturbance.

Much emphasis in the antenatal period is placed on anticipatory care, but proportionately more on the physical health of the mother and fetus than on the health of the wider family group. Involving fathers and sometimes siblings during pregnancy can pay dividends. Not only is there a chance of promoting better understanding of what is going on in a physical and emotional sense, but there is less chance of things going wrong in the family once the baby is born. Minor mood disturbance in the mother, changes in sexual desire, feelings of rejection amongst existing children, and any feelings husbands may have of taking second place in a wife's affections, can all be addressed. Simple explanations and reassurances that these are not uncommon feelings connected with pregnancy can stop them becoming magnified into family problems.

The same process can be continued in the early years of parenthood by actually involving fathers in issues which affect the parenting of children and by informing couples of the impact children can have on a marriage.

The possibility of collaborative work, particularly with health visitors, midwives, and visiting psychotherapists, extends the scope of this preventive work and allows more disturbance to be recognized at an early stage.

Opportunities for preventing the long-term harmful consequences of family break-up

As increasing numbers of families face separation and divorce, with over one fifth of involved children affected in ways that bring them into contact with their doctor, general practitioners have an obligation to identify and monitor disturbances related to family breakup. (Caplan, 1986).

By building up a supportive relationship with the family and by being prepared to see parents and children together as well as offering conciliation counselling, future disturbance can be ameliorated. It is reasonable to offer parents factual information concerning the risks to their children's emotional health. Children invariably want their parents to stay together, whatever the difficulties at home and however impossible it is. They continue to need contact with both parties. Easy access to the non-custodial parent is important, and should be encouraged. It should be stressed that parents need to co-parent, even if they choose not to live under the same roof. Moreover, they need to explain to their children that they are separating not because of any aspect of the children's behaviour, but because they as adults cannot get on with one another. Children often have fantasies of being instrumental in causing the rift between their parents and also frequently imagine they can bring about a reconciliation. These erroneous notions need to be addressed directly and the children told preferably by their parents, that, whatever the outcome, they will go on being loved and that they share none of the responsibility for the breakup. Children caught in marital conflict often have their needs pass unnoticed. It is bound to be disturbing, and behaviour which seems pathological to the parents is often a normal expression of the anguish the children feel. This again needs explaining.

SCREENING FOR FAMILY DISTURBANCE

This approach to the early detection of family problems is still in its infancy. One technique which has received support is the Family APGAR. First developed by Smilkstein in 1978 it takes the form of a five-item questionnaire given to patients which is designed to detect problems in family adaptation, partnership, growth, affection, and resolve (Table 14.1). Each question is scored 2, 1, or 0, depending on whether the person's answer is 'almost always', 'some of the time', or 'hardly ever'. Scores of 0 to 3 correlate with severe family dysfunction, 4 to 6 with moderate dysfunction, and 7 to 10 with little or no dysfunction. Its validity and reliability as a screening device have been tested, and though it measures a person's satisfaction with his or her family's functioning rather than the family functioning itself, it is a test which is useful and easy to adminster during the general practice consultation. It can also open the way to discussion about family problems. Almost 1 in 4 patients attending a family practice centre in America had evidence of family dysfunction using the device. Smilkstein himself found that 15 per cent of new patients scored 6 or less.

It seems likely that better screening devices will become available to general practitioners in the future, but the Family APGAR, combined

Table 14.1 *Family APGAR questionnaire*

	Almost always	Some of the time	Hardly ever
I am satisfied with the help that I receive from my family* when something is troubling me.	_______	_______	_______
I am satisfied with the way my family* discusses items of common interest and shares problem-solving with me.	_______	_______	_______
I find that my family* accepts my wishes to take on new activities or make changes in my life-style.	_______	_______	_______
I am satisfied with the way my family* expresses affection and responds to my feelings such as anger, sorrow, and love.	_______	_______	_______
I am satisfied with the amount of time my family* and I spend together.	_______	_______	_______

Scoring: The patient checks one of three choices which are scored as follows: 'Almost always' (2 points), 'Some of the time' (1 point), or 'Hardly ever' (0). The scores for each of the five questions are then totalled. A score of 7 to 10 suggests a highly functional family. A score of 4 to 6 suggests a moderately dysfunctional family. A score of 0 to 3 suggests a severely dysfunctional family.

* According to which member of the family is being interviewed the doctor may substitute for the word 'family' either spouse, significant other, parents, or children.

Source: Smilkstein, G. (1978). The family APGAR: a proposal for a family function test and its use by physicians. *Journal of Family Practice*, **6**, 1231–9.

with a greater awareness of the indirect forms of presentation of family disturbance, goes some way towards picking up problems early on.

A FAMILY LIFE FREE OF CONFLICT?

Whilst in the past the family afforded protection against a hostile world, binding people together out of necessity, it serves now to provide a milieu in which individuals can grow up to be emotionally satisified creatures. This is altogether a more difficult task, since, when emotional problems do arise in families, there is a tendency to blame someone for them: the husband blames the wife, the wife blames her parents, the parents blame the child, etc. The presupposition is that problems are avoidable rather than being an inherent part of family life.

The point of any preventive programme is not to eliminate family conflict, but to create a means by which conflict can be put to personal advantage. The challenge of being part of a family is in what one can make of that experience. The best definition of a happy family is not one that is free of problems, but one which, when something does go wrong, supports that person who carries most distress and does not blame others for its misfortunes. The general practitioner's job is to keep such ideals in mind and help the family to work towards them.

15 Teaching and learning

AIMS AND OBJECTIVES IN TEACHING

Training to be a general practitioner with an interest in family pathology is distinct from training to be a family therapist.

To become a family therapist takes a minimum of three years and involves both theoretical teaching and clinical supervision of at least 100 hours of direct work with families. Such training is provided by the Institute of Family Therapy in London. To become a doctor with an interest in family problems takes less time and relies mostly on a willingness to spend time with families and learn more of their problems. It does necessitate some reading. For the GP with an interest, introductory courses are available, whose aims—although limited—are to improve doctors' recognition, understanding, and management of family issues.

Certain objectives can be listed

1. To be able to identify family problems in the guise of physical and emotional symptoms.
2. To appreciate the role of the family in health and disease.
3. To be aware of the benefit of a family perspective over and above an individual one.
4. To know more about family structure and family life-cycle.
5. To be able to think in terms of 'systems'.
6. To be able to conduct an interview with a couple or family, so as to arrive at an assessment of what is going on in interpersonal terms.
7. To have clear objectives when it comes to actively working with families and to work towards agreement about them.
8. To respect every family's right to privacy.
9. To be prepared to intervene so as to promote change.
10. To know when and to whom to refer.

This implies new ways of thinking about families for most doctors.

TRADITIONAL VERSUS NEW THINKING

The traditional approach to clinical teaching is based on the individual, the patient who, though belonging to a family, is the sole object of care. Viewing instead the family as the object of care, capable of influencing the health of its individual members through an interactive process, offers a new perspective.

To assist in this new form of thinking, systems theory is of use. Teaching family-system concepts to doctors is made easier by the use of a familiar model. Christie-Seely (1981) uses a diagram of the human endocrine system with family members superimposed and teaches around four concepts central to systems theory, namely:

1 The whole system has to be understood in order to understand both the disturbance and the symptomatic individual: the whole being greater than the sum of the parts.
2 Homeostasis is essential to well-being and operates in sickness as well as health. It is maintained by complex positive and negative feedback mechanisms.
3 Emphasis needs to be placed not only on the parts themselves (the organs or the individual family members) but on the links or communication that exists between them. They provide the means by which the system's functioning can be properly understood.
4 Changes may occur in areas remote from the original focus. Change in just one organ or individual has the potential for affecting others some distance away.

Encouraging doctors to think along these lines and to apply systems—thinking to the families they encounter both demonstrates gaps in knowledge and provides an incentive to find out more about family structure and behaviour. Devices like the family wheel and the genogram can help to organize this information and provide convenient access to family problems. In that much family distress is conveyed to the doctor in physical terms, the use of a diagram which links physical to psychological states is useful when it comes to analysing the symptom-bearer (see Fig. 15.1).

FAMILY INTERVIEWING

Minuchin and Fishman (1981) suggest that only the person who masters techniques and then contrives to forget them, can become an expert family therapist. For the family doctor with more limited objectives, the goal is not so much to transcend technique as to become less conscious of method and more spontaneous in dealing with families. He can never hope to master a wide repertoire of interventions but can nevertheless be trained to use different aspects of himself in response to different social situations. He must show a willingness to learn through experiment and be prepared to be adventurous in his dealings with families. At the same time, he needs to be aware of his own limitations and the constraints imposed by time

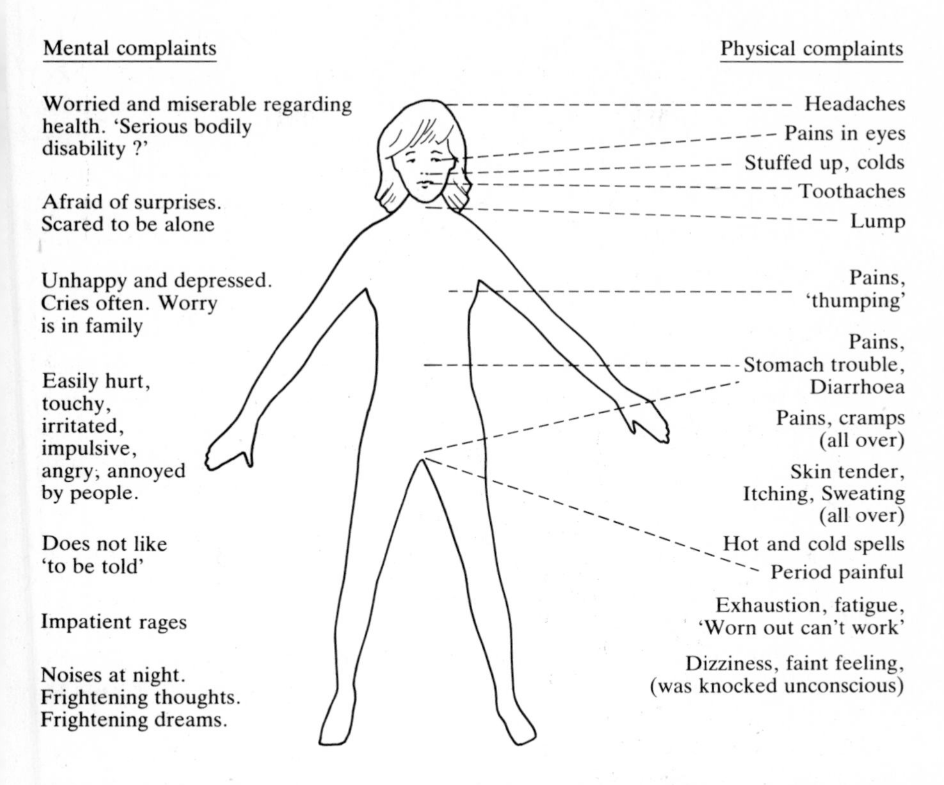

Fig. 15.1 The link between physical and psychological states. Foulkes, S. H. (1986) Group-Analytic Psychotherapy, Methods and Principles. Maresfield Library, London. Publ. H. Karnac (Books) Ltd. Copyright with Mrs Elizabeth Foulkes.

and confidentiality. Helping families is not all a matter of technique. When it comes to evaluating interventions, effectiveness relates more than anything to creating an atmosphere of trust and a climate in which change becomes a possibility. It is encouraging to know that more than two-thirds of families referred for family therapy respond to the application of simple methods, and this proportion should be higher in general practice where the population presenting with family problems is likely to have less intractable difficulties.

Learning in small groups has its advantages. Looking at oneself on video, engaging in role-play, and using sculpting techniques are activities appropriate to small groups of doctors interested in improving their skills.

Video analysis

For those doctors who are already accustomed to looking at their own consultations on video, analysing what goes on in sessions with families can be instructive. A video holds the attention of a small group in ways that an audiotape or verbal feedback cannot achieve, and gives additional non-verbal information which can be useful in forming an assessment. As with all discussion of videos in groups, it is important that the doctor who exposes his consultation to the others should have first opportunity to analyse his own performance. He should be encouraged initially to comment on what he considers to be good, and secondly on what he sees as weaknesses or missed opportunities. Next, his colleagues should express their opinions, taking care again to stress strengths in performance before pointing to deficiencies.

Role-play

To explore alternative ways of conducting a family interview, role-play is invaluable. The doctor may feel stuck with a problem, find it difficult to accept that other possibilities of management exist, or may simply have lost his sense of direction. By role-playing the situation, new insights may be given which lead to clearer objectives and new strategies. These can be used to advantage with the family.

Another use of role-play is in simulating consultations over hypothetical problems. Doctors are invited to become family members of their own chosing and a family system allowed to develop around a presenting problem. Such family simulations can give participants new confidence when confronted by similar problems in their own surgery.

Sculpting

To highlight issues such as interdependence, the exercise of power, and problems of emotional distance, family sculpting techniques are useful. The doctor in charge of a re-enacted family consultation invites his colleagues, who are acting the part of the family, to move into new positions in the room. By altering the physical distance between members—asking one to distance himself from the others, or to stand closer to the family group, instructing another to kneel or stand on a chair, or merely swapping around people's positions in the room, and then inviting them to share with the others some of their feelings about the positions they have adopted—everyone is helped to get in touch with underlying emotional responses within the family.

FURTHER TRAINING

Video-analysis, role-play, and sculpting, though desirable, are not essential to learning. As Skynner and Cleese (1983) point out, general practitioners who lack such learning opportunities can safely benefit from just looking more closely at the family systems of their patients. Additional help can come from attending courses, seminars, and study-days in various centres up and down the country. A list of such helping agencies forms an appendix to this Chapter.

The contribution that family therapists, marital therapists, and other commentators on family life can make to the understanding of family difficulties, cannot be over-emphasized. As a guide to further reading, I have added a selection of books which have been useful to me personally. I hope that they will be of interest to other doctors.

Appendix I: Helping Agencies

Institute of Family Therapy, 43, New Cavendish Street, London W1

A self-governing independent body formed in 1975, whose objectives are to:

1. Promote high standards in assessing and treating families.
2. Conduct research into family functioning.
3. Examine methods of intervention in families which have suffered bereavement, are undergoing disruption, or which include the physically or mentally-ill.
4. Look at family approaches to child sexual abuse.

It runs courses, workshops, and seminars throughout the year, designed for beginners as well as for professionals who work with families. It also provides a three-year training course in family therapy for people who have completed an introductory course at the Institute. There is an introductory course specifically designed for general practitioners. It houses a library of video-tapes on various aspects of family work which are available for hire.

The IFT also provides a clinical service charging fees on a sliding scale and accepting referrals from all sources including self-referrals.

Institute of Marital Studies, The Tavistock Centre, 120 Belsize Lane, London NW3

Founded in 1948 as the Family Discussion Bureau, the IMS is a unit of the Tavistock Institute of Human Relations, which, though not part of the National Health Service, is in the same building as the Tavistock Clinic.

The IMS has three connected functions:

1. To provide a therapeutic service to those experiencing difficulty in their marriage.
2. To offer training in marital and related family problems for workers in other settings.
3. To conduct research into the process of interaction within marriages and families, and between workers and their clients.

About one-third of available staff time is devoted to clinical work which provides the shared experiences necessary for training and research. Most referrals come from London and the Home Counties, although a proportion come from further afield. Fees are charged but most couples do not pay the full cost and are subsidized by a grant which the IMS receives from the Home Office.

The focus in working with couples is not so much on individual psychopathology as on the meaning and purpose of the transactions between the couples and on the creative aspects of conflict, and on those elements of pathology which point to new solutions. An example of the type of work carried out is provided by Christopher Clulow's book, *Marital therapy, an inside view*. (See Appendix II.)

National Marriage Guidance Council, Herbert Gray College, Little Church Street, Rugby, Warwickshire (Relate)

Counsellors work in approximately 150 local Marriage Guidance Councils which are autonomous in many respects. They are selected and trained centrally and receive regular supervision and support throughout their service.

Their main work is to provide counselling to individuals and couples who have difficulties in their marital or quasi-marital relationship. The aim is to help people work out their own solutions rather than have solutions imposed on them. For couples with a committed relationship who do not experience difficulties in other areas of their relationship, sex therapy may also be available.

The National Marriage Guidance Council also offers occasional short introductory courses open to doctors and other professional people about counselling aspects of their work. These courses have Section 63 approval. In some areas, counsellors work within the setting of a group practice helping general practitioners deal with the marriage and family problems encountered in everyday work.

Marriage guidance counsellors are unpaid. The running expenses and training costs are met only partly by grants from public funds, and most local marriage guidance counsellors therefore ask clients for contributions. No client is turned away because of inability to pay, however.

Scottish Institute of Human Relations, 56 Albany Street, Edinburgh

Provides training in family therapy and group work as well as personal analysis and individual child psychotherapy. Founded in 1969 it has become a major resource centre in Scotland for teaching about personal growth, human relationships, and institutional change using a psychodynamic model.

Family Policy Studies Centre, 231 Baker Street, London NW1

Established in 1983, as a bridge between research and policy, it aims to promote a positive dialogue between policy makers, academics, and practitioners. It produces a wide range of documents on family matters.

Marriage Research Centre, Central Middlesex Hospital, Acton Lane, London NW10

Set up in 1971 by Dr Jack Dominian to study marriage and divorce with a view to developing improved means of helping couples in difficulty as well as providing training and information for others. As well as publishing material on marital problems and divorce, it organizes courses and conferences for doctors, health visitors, and social workers.

Appendix II: Reading list

Texts

Family life, Graham Allan, 1985. Blackwell.
Secrets in the family, Lily Pincus and Christopher Dare, 1978. Faber.
Introduction to family therapy, Vincent D. Foley, 1986. Grune and Stratton.
One flesh: separate persons, Robin Skynner, 1976. Constable.
Families and how to survive them, Robin Skynner and John Cleese, 1983. Methuen.
Explorations with families: group analysis and family therapy, Robin Skynner. Ed. John R. Schlapobersky, 1987. Methuen.
Families and family therapy, Salvador Minuchin, 1974. Tavistock Publications Ltd.
Family therapy techniques, Salvador Minuchin and H. Charles Fishman, 1981. Harvard.
Family kaleidoscope, Salvador Minuchin, 1984. Harvard.
Problem solving therapy, Jay Haley, 1987. Jossey-Bass.
Strategic family therapy, Chloe Madanes, 1981. Jossey-Bass.
Basic family therapy, Philip Barker, 1986. Collins.
Marital tensions, Henry V. Dicks, 1967. Routledge and Kegan Paul.
Marital therapy: an inside view, Christopher F. Clulow, 1985. Aberdeen University Press.
Marital interaction and some illnesses in children, June Mainprice, 1974. Institute of Marital Studies.
Ethical issues in family medicine, Ronald J. Christie and C. Barry Hoffmaster, 1986. O.U.P.
Helping Families: systems, residential and agency responsibility, Peter Bruggen and Charles O'Brien, 1987. Faber and Faber.
Coping with disorder in the family, Ed. Jim Orford, 1987. Croom Helm.
Genograms in family assessment, Monica McGoldrick and Randy Gerson, 1985. W. W. Norton.
Children's problems: a parents guide to understanding and tackling them, Bryan Lask, 1985. Dunitz.
Life after marriage. A. Alvarez, 1982. Fontana.
The savage god: a study of suicide, A. Alvarez, 1974. Penguin.
The war over the family: capturing the middleground, Brigitte & Peter Berger, 1984. Penguin Books.
Sex therapy: a practical guide, K. Hawton, 1985. Oxford University Press.
Human sexual inadequacy, W. H. Masters & V. F. Johnston, 1970, Churchill Livingstone, London.

Other literature

The themes of personal loss, disturbed relationships, and individual growth and development provide novelists, biographers, and dramatists with some of their richest material. What follows is a brief selection of works that can be read and enjoyed, and at the same time, be instructive. All are available in paperback.

A death in the family. James Agee. Pan Books (Picador)
In the springtime of the year. Susan Hill. Penguin Books
Father and son. Edmund Gosse. Penguin Books
Knots. R. D. Laing. Penguin Books
Do you really love me? R. D. Laing. Penguin Books
The rainbow. D. H. Lawrence. Penguin Books
Tender is the night. F. Scott Fitzgerald. Penguin Books
Ghosts. Henrik Ibsen. Penguin Books
Le grand meaulnes. Alain-Fournier. Penguin Books
The war between the Tates. Alison Lurie. Penguin Books
The millstone. Margaret Drabble. Penguin Books
Portnoy's complaint. Philip Roth. Penguin Books
Franny and Zooey. J. D. Salinger. Penguin Books
Catcher in the Rye. J. D. Salinger. Penguin Books
A pale view of hills. Kazuo Ishiguro. Penguin Books
The bell jar. Sylvia Plath. Faber and Faber
Lord of the flies. William Golding. Faber and Faber
Eustace and Hilda trilogy. L. P. Hartley. Faber and Faber
The cocktail party. T. S. Eliot. Faber and Faber
The family reunion. T. S. Eliot. Faber and Faber
The unbearable lightness of being. Milan Kundera. Faber and Faber
Fierce attachments. Vivian Gornick. Virago Press
Manservant and maidservant. Ivy Compton-Burnett. Oxford University Press

Bibliography

Abrams, M. (1978). *Beyond three score and ten: a first report on a survey of the elderly*. Age Concern, Mitcham.

Allan, G. (1985) *Family life*. Basil Blackwell Ltd.

Anderson, R. (1987) The unremitting burden of carers. *British Medical Journal*, **294**, 73–4.

Bateson, G., Jackson, D.D., Haley, J., and Weakland, J. (1986) Towards a theory of schizophrenia. *Behavioral Science*, **1**, 251

Bax, M. (1978). Care of the handicapped child at home. *Clinics in developmental medicine,* No. 67, Chapter 6, 48–56.

Berger, B. and Berger, P. L. (1984). The war over the family; capturing the middle ground. Penguin, Harmandsworth M.

Brook, A. and Temperley, J. (1976). The contribution of a psychotherapist to general practice. *Journal of the Royal College of General Practitioners*, **26**, 86–94.

Brown, G.W., Birley, J. L. T., and Wing, J. K. (1972). The influence of family life on the course of schizophrenic disorders: a replication. *British Journal of Psychiatry,* **121**, 241–58.

Brown, G. W. and Harris, T. (1978). *Social origins of depression.* Tavistock Publications, London.

Caplan, G. (1961). *An approach to community mental health*. Grune and Stratton, Inc., New York.

Caplan, G. (1964). *Principles of preventive psychiatry*. Basic Books, New York.

Caplan, G. (1986) Preventing psychological disorders in children of divorce. *British Medical Journal* **292**, 1431–3, and 1563–6.

Carnwarth, T. C. M. and Johnson, D. A. W. (1987). Psychiatric morbidity among spouses of patients with stroke. *British Medical Journal*, **294**, 409–11.

Christie-Seely, J. (1981). Teaching the family system concept in family medicine. *Journal of Family Practice,* **13**, 391–401.

Clulow, C. (1984). Sexual dysfunction and interpersonal stress: the significance of the presenting complaint in seeking and engaging help *British Journal of Medical Psychology,* **57**, 371–80.

Crisp, A. H., Harding, B., and McGuinness, B. (1974). Anorexia nervosa: psychoneurotic characteristics of parents: relationship to prognosis. A quantitative study. *Journal of Psychosomatic Research,* **18**, 167–73.

Dare, C. (1979). Psychoanalysis and systems in family therapy. *Journal of Family Therapy,* **1**, 137–51.

Dawes, D. (1985). Standing next to the weighing scale. *Journal of Child Psychotherapy,* Vol. 11, No. 2. 77–85.

Deparment of Health and Social Security. (1974). *Report of the committee on one-parent families* (Finer report). 2 vols. London: HMSO.

Duvall, E. M. (1962). *Marriage and family development.* 5th Ed. Harper & Row, Publishers Inc., New York.

Edgell, S. (1980). *Middle-class couples*. London: Allen and Unwin.

Foley, V. D. (1986). *An introduction to family therapy*. 2nd Ed. Grune and Stratton Inc. Orlando, Florida, USA.

Gath, A. (1978). *Down's syndrome and the family: the early years.* Academic Press, London.

Goldberg, D. (1982). The recognition of psychological illness by general practitioners. *Psychiatry and general practice* (ed. A. W. Clare, and M. Lader). Academic Press, London.

Goldstein, M. J. and Rodnick, E. J. (1978). The family's contribution to the etiology of schizophrenia: current status. *Schizophrenia Bulletin,* **14**, 48–73.

Goodwin, D. W. *et al.* (1973). Alcohol problems in adoptees raised apart from alcoholic biological parents. *Archives of General Psychiatry,* **28**, 238–43.

Graham, H. and Sher, M. (1976). Social work and general practice. *Journal of the Royal College of General Practitioners,* **26**, 95–105.

Helsing, K. J., Szklo, M., and Cornstock, G. W. (1981). Factors associated with mortality after widowhood. *American Journal of Public Health,* **17**, 802–9.

Hetherington, E. M. *et al.* (1982). *Effects of divorce on parents and children, in non-traditional families* (ed. Lamb, M. E.), pp. 233–88. Lawrence Erlbaum, Hillsdale, New Jersey, USA.

Hoare, P. (1984). The development of psychiatric disorder among school children with epilepsy. *Developmental Medicine and Child Neurology,* **26**, 3–13.

Holmes, T. H. and Rahe, R. H. (1967). The social re-adjustment rating scale. *Journal of Psychosomatic Research,* **11**, 213–18.

Johnston, I. D. A., Hill, M., Anderson, H. R., and Lambert, H. P. (1985). Impact of whooping cough on patients and their families. *British Medical Journal,* **290**, 1636–8.

Kalucy, R. C., Crisp, A. H. and Harding, B. (1977). A study of 56 families with anorexia nervosa. *British Journal of Medical Psychology,* **50**, 381–95.

Kety, S. S. (1980). The syndrome of schizophrenia: unresolved questions and opportunities for research. *British Journal of Psychiatry,* **136**, 421–36.

Kiernan, K. (1983). The structure of families today: continuity or change? *OPCS Occasional Paper,* No. 31, 17–36

Kraus, A. S. and Lilienfeld, A. M. (1959). Some epidemiologic aspects of the high morality rate in the young widowed group. *Journal of Chronic Diseases,* **10**, 207–17.

Lask, B. and Kirk, M. (1979). Childhood asthma: family therapy as an adjunct to routine management. *Journal of Family Therapy,* **1**, 33–49.

Leventhal, J. M., Bentovim, A., Elton, A., Tranter, M., and Read, L. (1987). What to ask when sexual abuse is suspected. *Archives of Disease in Childhood,* **62,** 1188–95.

Lindemann, E. (1944). Symptomatology and management of acute grief. *American Journal of Psychiatry,* **101,** 141–8

MacKeith, R. (1973). The feelings and behaviour of parents of handicapped children. *Developmental Medicine and Child Neurology,* **15**, 524–5.

Mayou, R., Foster, A., and Williamson, B. (1978). The psychological and social effects of myocardial infarction on wives. *British Medical Journal,* **1**, 699–701.

Meyer, R. J. and Haggerty, R. J. (1962). Streptococcal infections in families. *Pediatrics,* **29**: 539–49.

Minuchin, S. (1974). *Families and family therapy*. London, Tavistock.

Minuchin, S. (1984). *Family Kaleidoscope.* Harvard University Press.

Minuchin, S. And Fishman, H. C. (1981). Family therapy techniques. *Harvard University Press.*

Minuchin, S., Rosman, B. L., and Baker, L. (1978). *Psychosomatic families: anorexia nervosa in context.* Harvard University Press, Cambridge, Mass., USA.

Parkes, C. M., Benjamin, B., and Fitzgerald, R. G. (1969) Broken heart: a statistical study of increased mortality amongst widowers. *British Medical Journal,* **1**, 740–3.

Pearson, J. *The ultimate family—the making of the Royal House of Windsor.* (1987) Grafton Books.

Pincus, L. and Dane, C. (1978). Secrets in the family. Faber & Faber Ltd, London.

Querido, A. (1963). *The efficiency of medical care.* H. E. Stenfert Kroese (N. V.) Leiden.

Pollock, G. H. (1970). Anniversary reactions, trauma and mourning, *Psychoanalytic Quarterly,* **39**, 347–71.

Reading, P. (1986). A study of the impact of mental disturbance on the family as seen by key relatives. *Mind.* Oxford.

Richards, M. (1985). Paper given to The Marriage Research Centre's 1985 Conference on 'Family stress and children'.

Rosenberg, E. E. and Pless, I. B. (1985). Clinician's knowledge about the families of their parents. *Family Practice,* Vol. 2, No. 1., 23–8.

Rutter, M. (1975), *Helping troubled children.* Penguin Books Ltd.

Schleifer, S. J., Keller, S. E., Camerino, M., Thornton, J. C., and Stein, M. (1983). Suppression of lymphocyte stimulation following bereavement. *J.A.M.A.* **250**, 374–7.

Skynner, R. and Cleese, J. (1983). Families and how to survive them. Methuen, London Ltd.

Smilkstein, G. (1978). The family APGAR: a proposal for a family function test and its use by physicians. *Journal of Family Practice,* **6**, 1231–9.

Szasz, T. S., (1961). *The myth of mental illness.* New York (Hoeber-Harper).

Tomson, P. R. V. (1985). Genograms in general practice. *Journal of the Royal Society of Medicine Suppl.* No. 8, Vol. 78., pp. 34–9.

Vaillant, G. (1983). *The natural history of alcoholism. Causes, patterns and paths to recovery.* Harvard University Press, Cambridge, Mass., USA.

van Eerdewegh, M. M. *et al.* (1985). The bereaved child: variables influencing early psychopathy. *British Journal of Psychiatry,* **147**, 188–92.

von Bertalanffy, L. (1968). *General Systems Theory: foundations, development, application.* Braziller: New York.

Wenger, G. C. (1984). *The supportive network: coping with old age. National Institute Social Services Library, No. 6.* London: George Allen and Unwin.

White, M. (1983). *Long-term unemployment and labour markets.* Policy Studies Institute, London.

Williams, P. R. (1982). Recording symptoms and family relationships: a proposal. *British Medical Journal,* **284**, 1919.

Williams, P. R., Argent, E. H. M., and Chalmers, C. (1981). A study of an urban health centre: factors influencing contact with mothers and their babies. *Child: Care, Health and Development,* **7**, 255–66.

Wilmott, P. (1986). *Social network: informal care and public policy. Research Report 655.* Policy Studies Institute, London.

Yarrow, M. R., Clawen, J. A., and Robbins, P. R. (1955). The social meaning of mental illness. *Journal of Social Issues,* **X1 (4)**, 33–48.

Young, M. and Willmott, P. (1973). *The symmetrical family.* Routledge and Kegan Paul, London.

Index